Radioactive Waste Disposal at Sea

Global Environmental Accord: Strategies for Sustainability and Institutional Innovation
Nazli Choucri, editor

Peter M. Haas, Robert O. Keohane, and Marc A. Levy, editors, *Global Accord: Environmental Challenges and International Responses*

Ronald B. Mitchell, *Intentional Oil Pollution at Sea: Environmental Policy and Treaty Compliance*

Robert O. Keohane and Marc A. Levy, editors, *Institutions for Environmental Aid: Pitfalls and Promise*

Oran R. Young, editor, *Global Governance: Drawing Insights from the Environmental Experience*

Jonathan A. Fox and L. David Brown, *The Struggle for Accountability: The World Bank, NGOs, and Grassroots Movements*

David G. Victor, Kal Raustiala, and Eugene Skolnikoff, editors, *The Implementation and Effectiveness of International Environmental Commitments: Theory and Practice*

Mostafa K. Tolba with Iwona Rummel-Bulska, *Global Environmental Diplomacy: Negotiating Environmental Agreements for the World, 1973–1992*

Karen T. Litfin, editor, *The Greening of Sovereignty in World Politics*

Edith Brown Weiss and Harold K. Jacobson, editors, *Engaging Countries: Strengthening Compliance with International Environmental Accords*

Oran R. Young, editor, *The Effectiveness of International Environmental Regimes: Causal Connections and Behavioral Mechanisms*

Ronie Garcia-Johnson, *Exporting Environmentalism: U.S. Multinational Chemical Corporations in Brazil and Mexico*

Lasse Ringius, *Radioactive Waste Disposal at Sea: Public Ideas, Transnational Policy Entrepreneurs, and Environmental Regimes*

Radioactive Waste Disposal at Sea
Public Ideas, Transnational Policy Entrepreneurs, and Environmental Regimes

Lasse Ringius

The MIT Press
Cambridge, Massachusetts
London, England

© 2001 Massachusetts Institute of Technology

All rights reserved. No part of this book may be reproduced in any form by any electronic or mechanical means (including photocopying, recording, or information storage and retrieval) without permission in writing from the publisher.

Set in Sabon by The MIT Press.
Printed and bound in the United States of America.

Library of Congress Cataloging-in-Publication Data

Ringius, Lasse.
Radioactive waste disposal at sea : public ideas, transnational policy entrepreneurs, and environmental regimes / Lasse Ringius.
p. cm. — (Global environmental accord)
Includes bibliographical references and index.
ISBN 0-262-18202-5 (hc. : alk. paper) — ISBN 0-262-68118-8 (pbk. : alk. paper)
1. Radioactive waste disposal in the ocean—International cooperation. I. Title. II. Global environmental accords.
TD898.4 .R56 2000
363.72'89—dc21   00-031879

Contents

Series Foreword   vii
Preface   ix
1 Introduction   1
2 History of Efforts to Control Ocean Disposal of Low-Level Radioactive Waste   21
3 Transnational Coalitions of Policy Entrepreneurs, Regime Analysis, and Environmental Regimes   35
4 Scientific Advice and Ocean Dumping: Knowledge-Based Regime Analysis   59
5 Ocean Dumping and U.S Domestic Politics: Power-Based Regime Analysis   73
6 Negotiating the Global Ocean Dumping Regime: Interest-Based Regime Analysis   87
7 Explaining Regime Formation   109
8 Changing the Global Ocean Dumping Regime   131
9 Explaining Regime Change   153
10 Conclusion   171
Key Events   195
Notes   197
References   231
List of Interviews Cited   255
Index   257

Series Foreword

A new recognition of profound interconnections between social and natural systems is challenging conventional constructs and the policy predispositions informed by them. Our current intellectual challenge is to develop the analytical and theoretical underpinnings of an understanding of the relationship between the social and the natural systems. Our policy challenge is to identify and implement effective decision-making approaches to managing the global environment.

The series on Global Environmental Accord adopts an integrated perspective on national, international, cross-border, and cross-jurisdictional problems, priorities, and purposes. It examines the sources and the consequences of social transactions as these relate to environmental conditions and concerns. Our goal is to make a contribution to both intellectual and policy endeavors.

Nazli Choucri

Preface

My first encounter with international regimes took place in the late 1980s. At that point I was employed by the Danish Ministry of Foreign Affairs, and one of my tasks was writing memos and reports on international commodity arrangements for jute, tropical timber, tin, and cocoa. I have since then been increasingly interested in the issue of international cooperation, particularly cooperation for the protection of the environment and natural resources.

Writing this book offered me the opportunity to focus on an important environmental issue—disposal of radioactive waste at sea—and a fascinating international environmental regime—the global ocean dumping regime. The book is based on my doctoral dissertation, completed at the Department of Political and Social Sciences at the European University Institute (EUI) in Florence, in which I examined prominent regime theories and presented an empirical study of the global ocean dumping regime and regulation of ocean disposal of low-level nuclear waste. In this book I have further developed the theoretical argument outlined in the dissertation, in particular the global significance of public ideas, transnational coalitions of policy entrepreneurs, and environmental nongovernmental organizations.

At EUI, Giandomenico Majone and the late Susan Strange introduced me to the role of ideas and information in policy analysis and regime analysis. A substantial part of my empirical research was carried out during a two-year stay as a visiting researcher at the Center for International Studies and in the Department of Urban Studies and Planning at the Massachusetts Institute of Technology. Eugene Skolnikoff and Larry Susskind gave invaluable advice and enthusiastically supported me during my stay in Cambridge. I have revised my dissertation over a longer period in which I have been focusing primarily on the issue of global climate change at the Center for

International Climate and Environmental Research (CICERO) in Oslo. I have benefited much from presentations and discussions of my empirical findings and theoretical argument in Florence, Cambridge, and Oslo. The book has benefited greatly from the comments and suggestions of two anonymous reviewers and one non-anonymous reviewer, Edward L. Miles. I have received useful comments and suggestions for improvements from Giandomenico Majone, Olav Schram Stokke, Arild Underdal, and Jørgen Wettestad. And Nazli Choucri, Clay Morgan, and Paul Bethge helped me navigate the waters of the academic publishing world.

Others also helped significantly. I am particularly grateful to Jesper Grolin, who provided me with key documents and much-needed encouragement at a crucial early point. Over the years, officials in the Environmental Protection Agency in Copenhagen, in particular Kjeld F. Jørgensen and Bente Mortensen, never failed to answer my questions and inquiries, and the secretariat at the International Maritime Organization kindly allowed me to observe the fourteenth consultative meeting of the parties to the London Convention in London in 1991. I also wish to thank the many interviewees who took time to share their knowledge and expertise with me. I gratefully acknowledge the financial support received from the Danish Social Science Research Council, the Danish Research Academy, the Fulbright Commission, Consul Axel Nielsen's Foundation, the Denmark-America Foundation, EUI, and CICERO.

Portions of chapters 8 and 9 were originally published in different form in my article "Environmental NGOs and Regime Change: The Case of Ocean Dumping of Radioactive Waste" (*European Journal of International Relations* 3 (1997), no. 1: 61–104; © 1997 Sage Publications).

I am grateful to my parents, John Ringius and Hanne Duetoft, who stimulated my early interest in politics and have continuously been fine sources of intellectual inspiration and support. I am finally indebted to Jill Ringius, without whose support and patience this book probably would not have been written. Because of her many fine suggestions for improvements, the book is both more readable and more coherent than it otherwise would have been.

Radioactive Waste Disposal at Sea

1
Introduction

In this book I seek to narrow the ideational and normative gap in studies of environmental regimes. I intend to demonstrate that we can improve our understanding of the dynamics of international environmental regimes considerably by focusing in a systematic way on ideas and ideational factors. In particular, I document that powerful public ideas and policy entrepreneurs can significantly influence the processes of regime formation and regime change.

Public ideas are widely accepted ideas about the nature of a societal problem and about the best way to solve it.[1] They are about the welfare of society, and therefore they differ from private concerns. By defining how societal problems are perceived, public ideas shape policy and public debate about policy. Ideas about which there is societal consensus remain stable over time, and they differ from ideas that are altered frequently and from ideas that are promoted by special-interest groups.[2] Public ideas are held by society, not just by individuals, and therefore there is a fundamental difference between public ideas and individual beliefs.[3] Under certain circumstances, the interplay of public ideas and transnational[4] coalitions of policy entrepreneurs creates and changes environmental regimes.

Studies of regime building and regime change in the environmental field have been concerned predominantly with the use of power, the pursuit of rational self-interest, and the influence of scientific knowledge. Scholars have paid far less attention to the role of ideational factors in creating and changing the principles and norms of environmental regimes. Surprisingly, the processes by which the norms and principles of regimes are developed and changed have received little explicit attention, despite widespread agreement among regime analysts that regimes are defined and constituted by their underlying norms and principles as well as by their rules and decision-making procedures. Although a growing number of studies of public policy and of international relations ascribe great significance to ideas and norms, analysts of environmental regimes pay little attention to ideas and norms, and their development and significance have not been empirically documented. Moreover, this neglect of the ideational and normative aspects of environmental regimes stands in marked contrast to the importance that politicians, national and international environmental authorities, and environmentalists ascribe to awareness raising, campaigning, and education in improving environmental protection.

Since the early 1960s, states have signed numerous international environmental treaties and agreements and have been building environmental regimes. Initially, regional marine pollution and regional air pollution attracted most of the political attention that was paid to environmental problems. A number of regional regimes were created to deal with ocean dumping, with land-based sources of marine pollution, with acid rain, and with other environmental issues. In the 1980s and the 1990s, protection of the stratospheric ozone layer, preservation of biological diversity, and prevention of global climate change appeared on the international environmental agenda, and regimes were created to address such global environmental issues. By 1992, there were 170 international environmental agreements and treaties (UNEP 1997).[5] States will probably continue to build regimes and to develop international regulatory machinery for environmental protection for the foreseeable future.

There is no overarching international authority for the protection of the environment. Nonetheless, regimes may be able to increase cooperation among countries. According to Krasner (1983, p. 2), "regimes can be defined as sets of implicit and explicit principles, norms, rules, and decision-making procedures around which actors' expectations converge in a given area of international relations."[6]

How are global environmental regimes created, and by whom? Once created, how and by whom are they changed? At first, analysts mostly paid attention to why and how regimes are initiated and built; less attention was paid to why and how regimes change and transform. More recent studies have made greater efforts to analyze the effectiveness of regimes in changing social behavior and ultimately solving environmental problems. These studies have started from the assumption that regimes matter, and they have aimed to identify causal pathways.[7] Despite many insightful analyses, there are still important lessons to be learned about the establishment and the transformation of environmental regimes.

Focusing on the global ocean dumping regime and on the developments in the control of ocean dumping of low-level radioactive waste (commonly referred to as *radwaste disposal*) that have taken place under that regime, I examine the formation and change of the deep normative structure within which global ocean dumping policy—particularly with regard to radwaste disposal—has been embedded. The case of the global ocean dumping regime and radwaste disposal demonstrates the significance of transnational coalitions of policy entrepreneurs and public ideas in a powerful way. In particular, it offers an opportunity to examine more carefully the political construction of a global environmental problem, the role of persuasion and communication in an international setting, and the formation of international public opinion. These themes have not previously been developed and integrated in regime analysis.

I do not claim in this book that the influence of ideas (or, as some prefer, the power of ideas) alone can explain regime development. Policy development should not be reduced to just a question of ideas; it is important to understand how ideas and interests together influence policy. Scholars who suggest that ideas have a significant independent impact on policy have not always been sufficiently careful or precise when specifying under what conditions ideas are likely to be influential, why one set of ideas had more force than another in a given situation, and in what way a particular idea made a difference. To increase our understanding of the policy influence of ideas, we therefore need to focus more systematically on key issues: What kinds of ideas matter? In what way do they influence behavior? In what ways and by whom are influential ideas transmitted? Through careful examination of these important issues, the ideational approach to political analysis improves our understanding of policy initiation and policy development at the national and the international level. The analytical ambition of regime analysis should be to integrate ideas and interests, rather than to segregate them. By focusing on the role of ideas, it seems possible to develop auxiliary hypotheses that could supplement better-established theories about how power and interests influence regimes.

In this book I hope to show that more attention should be given to how the interplay of public ideas and policy entrepreneurs influences environmental regimes. Scholars should carefully distinguish the different paths to regime establishment, or the distinct processes by which regimes are built, and should identify the theoretical models that best explain the different paths to regime establishment. I explore a distinctively ideational type of regime dynamics. Moreover, I carefully compare power-based, interest-based, and knowledge-based theories of regimes, and I show that these theories, despite their many valuable insights, explain neither the formation nor the change of the global ocean dumping regime satisfactorily. In particular, prominent regime theories overlook the significance of public ideas and policy entrepreneurs.[8]

Ideas and policy entrepreneurs are combined in complex ways in domestic and international environmental politics. It is well known that environmentalists, ecologists, scientists, experts, and the mass media often can play significant roles when environmental problems are discovered and needs for corrective environmental policy are identified. However, it has largely gone unnoticed, as my study shows, that policy entrepreneurs active inside governments, government agencies, and international organizations also are actively involved in discovering global environmental problems, constructing and framing global environmental issues, and building environmental regimes. In other words, politicians, public administrators, and international bureaucrats are not only responding to the pressures of interest groups or just passively following the swings of public opinion. Policy makers and public administrators may actively discover new societal needs, create public value, and provide environmental protection and other national and international public goods.[9] In this book I document that transnational coalitions of policy entrepreneurs can play a major role in building a regime for the protection of the global environment.

I also document and analyze a significant process of regime change—in other words, a fundamental change of a regime's principles and norms—that resulted in a major change in international environmental policy.[10] Those who expect regimes to have no significant impact on state behavior would doubt the policy consequence of a regime change. I document that global environmental nongovernmental organizations (ENGOs) can act as catalysts for regime change, even when scientists and powerful states support existing policies. In view of the strong opposition to regime change by Britain, France, Japan, and the United States and the absence of a scientific consensus on environmental and human health risks, the regime change that I document contradicts not only traditional power-based theories that predict that regime change will only occur in response to changes in the power and interests of dominant states but also more recent knowledge-based theories arguing for the power of epistemic communities. Though it is true that states are the only formally recognized members of regimes, ENGOs are often de facto members with considerable influence. Finally, the lack of scientific evidence of damage to humans and the environment from ocean dumping of radwaste raises an important question: Does this recent regime development constitute "good"[11] and effective or unwise and negative global environmental policy?

The London Convention and Radwaste Disposal

In this book I investigate the formation and change of the global regime regulating ocean dumping of wastes, a regime established by the so-called London Convention of 1972.[12] I examine in detail the events that resulted in the establishment of this environmental regime. I also examine the developments in the control of radwaste disposal that have taken place under the regime.

In 1946 a number of countries began unilaterally disposing of low-level radioactive waste at sea. Other countries, though concerned about the environmental effects of such dumping, failed to halt it. Radwaste disposal continued after the establishment of the global ocean dumping regime in 1972. Hence, the regime initially permitted and regulated controlled radwaste disposal, and Britain and the United States (and later France and Japan) strongly supported this disposal practice. A moratorium temporarily halting radwaste disposal was imposed in 1983, however, and no official dumping has taken place since then. Agreement on a global ban on radwaste disposal was reached in 1993. According to an advisor to Greenpeace International, the decision to prohibit radwaste disposal was "a major step forward by the world community in making a commitment to protect the world's oceans."[13] The 1993 ban is evidence of a recent dramatic change of the regime—a change from permissive allowance of radwaste disposal to an emphasis on precaution and prevention.

Environmental regimes are manifestations of interstate and transnational cooperation. Keohane (1984, p. 51) has defined cooperation as follows: "Cooperation occurs when actors adjust their behavior to the actual or anticipated preferences of others, through a process of policy coordination." In this book I examine the global ocean dumping regime and radwaste disposal regulation as a case of global environmental regime formation and regime change. The formation and the transformation of the global ocean dumping regime—the two most important institutional developments with respect to global policy for radwaste disposal—are examined in detail.

Examining policy development requires a time perspective of at least 10 years.[14] In this book I present a longitudinal study of more than 20 years of policy and institutional development within one environmental issue area. The primary focus is on political and ideational issues, although attention is paid to the technical and scientific aspects of ocean dumping and radioactive waste regulation.

International Regimes and Ideas

A generally accepted comprehensive theory of how regimes form and how they change has yet to be developed. Scholars have only recently begun to conduct detailed and carefully designed empirical studies of environmental regimes intended to evaluate alternative regime theories. In their careful review of power-based and alternative theories of regimes, Haggard and Simmons (1987, p. 502) observed that "hegemonic interpretations of regimes are not always clear about what hegemons actually *do* to promulgate and maintain a given set of rules." Reviewing a collection of prominent regime studies, Young (1986, p. 110) concluded pessimistically that "confusion abounds when we turn to the actual processes through which institutions emerge." Young (1983, pp. 93–113) suggested distinguishing in an analytical sense among self-generated, negotiated, and imposed regimes, and among three corresponding processes resulting in establishment of regimes.

Regime scholars distinguish three different research strategies or sorts of hypotheses in their analyses of regime formation: *power-based*, *interest-based*, and *knowledge-based*.[15] Put briefly, power-based hypotheses deal with the role of power in a material and military sense, interest-based hypotheses are concerned with the role of interests (particularly the pursuit of self-interest), and knowledge-based hypotheses focus on the role of knowledge and perception in international politics. An individual scholar may not fit easily into any one of these three approaches, and some scholars contribute to more than one theoretical approach.

Power theorists, most prominently realists, start from an assumption about the anarchic nature of the international state system and claim that ideational factors are epiphenomenal because states fundamentally strive to expand their power and to protect themselves against other states striving for power and wealth. Supporters of power-based hypotheses point to the distribution of power capabilities among states and expect shifts in power capabilities to result in regime change. This view is closely associated with the claim that a regime is generally created by a hegemon that possesses "a preponderance of material resources" (Keohane 1984, p. 32). It is indeed relevant to examine power-based hypotheses; as I will show, the United States acted as an international leader when the global ocean dumping regime was built.

Neoliberals stress the significance of interests and assume that states make their choices independently in order to maximize their own returns. They claim that cooperation is based largely on self-interest and on realization of joint gains. Power politics has to give way to bargaining, compromise, and cooperation, and nonstate actors (especially scientists and international organizations) may matter. Interest-based explanations often stress that states' incentives to cooperate vary with the nature of the problems they hope to solve. For instance, in regard to the possibilities for successfully building regimes, some distinguish between benign problems (coordination problems and conditions for realizing benefits of coordination) and malign problems (where a skewed distribution of incentives to cooperate give rise to distributive conflict) (Underdal 1987). Neoliberals have identified a limited number of leadership types that, under certain conditions, are instrumental when self-interested states attempt to build regimes (Young 1991). Scholars who stress the importance of interests rarely focus systematically on the significance of ideas and perception in regime development.

A third group of scholars—mainly reflectivists, cognitivists, and social constructivists—are impressed with the significance of perception and cognition when states build regimes.[16] Because states' interests are perceived and are outcomes of cognitive processes, these scholars claim, they cannot be determined objectively and deductively beforehand. In their view, knowledge and values necessarily influence how states define their interests and formulate their preferences. The knowledge-based approach to regime analysis examines scientific and technical knowledge as a source of power and is concerned with how scientific knowledge can induce states to initiate the creation of regimes. The claim that technical expertise and scientific knowledge play prominent roles in international environmental politics has, more recently, led to the notion that "epistemic communities" (transnational networks of like-minded experts) significantly influence international policy. Peter Haas stresses the importance of consensual knowledge and causal ideas (assumptions about cause-effect relations and how the world works) in his explanation of regional efforts to protect the Mediterranean Sea (1990a). Karen Litfin, in her study of cooperation to protect the stratospheric ozone layer, finds that "knowledge brokers" ("intermediaries between the original researchers, or the producers of knowledge, and the policy makers who consume that knowledge but lack the time and training necessary to absorb the original research") were very influential in the regime-building process (1994, p. 4). Haas is primarily concerned with relationships between scientists and policy makers, while Litfin is more concerned with relationships between science and politics. Despite their differences, both Haas and Litfin focus on the power of scientific knowledge in environmental regime processes.

Studies of the regulation of radwaste disposal under the global ocean dumping regime have pointed to the special influence of scientists and have emphasized the significance of scientific and trans-scientific issues.[17] The trans-scientific issues are a particular group of public policy issues that, although partly concerned with issues of science and technology, cannot be resolved by science (Weinberg 1972). But my study shows that regulation of radwaste disposal under this environmental regime is not satisfactorily explained by pointing to the influence of scientists or to trans-scientific issues.

It is quite clear, although international environmental politics is strongly influenced by normative beliefs and by widely held perceptions of the health of the environment, that scholars of environmental regimes have paid little systematic attention to public ideas, policy entrepreneurs, and broader public perceptions. Moreover, in explaining cooperation on environmental protection, few if any contributors to regime analysis have ascribed much significance to the creation of legitimacy, to the justification of policy, or to the mobilization of public opinion. Stated differently: It seems clear that many regime analysts would concur with the view that "the politics of new resource issues ["those resources that are of recent origin elsewhere than on or under land: in ocean beds, outer space, climate and the geosynchronous orbit"] is unlikely to be sustained by the hue and cry of mass publics or burdened or facilitated by their mobilization and intervention" (Rosenau 1993, p. 81).

The inattention to public ideas and ideational factors reflects the fact that regime analysis (including knowledge-based regime analysis to some extent) adopts basic concepts and notions of realism, rational-choice theory, game theory, and the study of regional integration and international organizations. Scholars in these traditions have focused almost exclusively on strategic action and on rationalistic elements of elite behavior within and among states and international organizations. Although evidently important, power understood in a materialistic sense and distributive aspects of environmental policy within and among states have attracted inordinate attention as a result. Scholars ignore the roles of policy entrepreneurs and public ideas in the mobilization of political and public support, in the justification of policy, and in the establishment of political legitimacy in regime processes.

In contrast, a number of recent studies in comparative politics and international relations have concluded that ideas constitute a significant independent variable in policy processes. Supporters of the emerging ideational approach have found evidence of causal effects of ideas across a number of areas of foreign policy and international policy. These studies, which cover international cooperation in public health, macroeconomic policy, monetary policy, trade policy, development strategy, human rights, military affairs, foreign policy, international political change, and European Union policy making, conclude that policy choices and policy development can be explained satisfactorily only if ideational factors are taken into account.[18] Regime analysts, however, apart from studying the role of scientists, epistemic communities, and knowledge brokers, have considered ideational factors to be of minor significance for environmental regimes.

In addition to presenting an empirical study of ideational factors, I empirically document a global regime process that differs significantly from exclusively power-based, interest-based, and knowledge-based propositions about the dynamics of regimes. Careful empirical studies are necessary in order to better understand the impact of ideas on international policy, in particular empirical studies that pay closer attention to domestic politics and broaden the narrow focus on state and policy elites prevalent in regime analysis as well as in many studies of ideas-based policy change.[19] The ideational approach is an analytically fruitful way of connecting policy entrepreneurs, ideas, and the public arena. Also, regime analysts should acknowledge the significance of political legitimacy, policy entrepreneurs identifying common goals, and the concern of the mass public. Moreover, I pay considerable attention to the role of political interests in shaping simple, intuitively appealing metaphors, images, and symbols that influence regimes.

Principal Findings and Argument

My detailed empirical study of the formation and the change of the global ocean dumping regime shows that this regime was established in response to new environmental ideas, beliefs, and values rather than in response to concerns for protection of more tangible economic and political interests of states. Ideational hypotheses capture important stages, processes, and actors in the formation and change of this global environmental regime. I generalize an ideational path of regime formation by identifying the ideational conditions under which regime formation took place and by examining the responsible causal processes and the prime actors.[20] I also demonstrate that power-based, interest-based, and knowledge-based analyses give an unsatisfactory account of the formation and transformation of this regime.

The formation of the global ocean dumping regime is best described as essentially a two-step process. First, in the late 1960s and the early 1970s, the idea and metaphor that "the oceans are dying" gained tremendous political influence among important policy makers, particularly in the United States. The core of this idea was that the oceans generally were polluted and that this pollution was first and foremost the result of the industrial activity that characterizes modern society. Being an extremely powerful public idea and metaphor, the "dying oceans" idea framed ocean dumping as a global environmental problem of some urgency and importance. The United States quickly responded to this emerging public idea by establishing domestic ocean dumping regulation and by playing a leadership role in the construction of the global ocean dumping regime. In the second step, a broad range of nonstate actors, including prominent individual scientists and ecologists, specialized organizations of the United Nations, and the international mass media, framed the issue of ocean dumping, placed it on the global agenda, and mobilized international political and public support for the global ocean dumping regime. As a result, the "dying oceans" idea became institutionalized in the form of the global ocean dumping regime.

More analytically, my case study shows that a loosely organized transnational coalition of policy entrepreneurs, consisting primarily of environmentalists, politicians, and national and international government officials, is able to build an environmental regime by constructing a seemingly evident and indisputable need for environmental cooperation. The influence of transnational coalitions of policy entrepreneurs stems from their ability to politically construct international environmental problems and to act as international norm setters. In brief, they control the process of international problem definition. To attract political attention, mobilize international public opinion, and build public pressure, and at the same time delegitimize competing problem definitions and neutralize blocking coalitions, transnational coalitions of policy entrepreneurs use global channels of mass communication to spread public ideas, metaphors, and slogans that dramatically and powerfully capture targeted environmental problems. Using scientific evidence of environmental damage to persuade policy makers is not a necessary component of these coalitions' strategy for influence. Transnational coalitions of policy entrepreneurs attempt to create fair and flexible institutional solutions that will increase the participation of governments in regimes.

Persuasive and credible demonstration of a need for international environmental protection is critical in the problem-definition and agenda-formation stages of regime formation. Powerful public ideas and symbols of environmental destruction are necessary to focus public and political attention on global environmental problems. Transnational coalitions of policy entrepreneurs deliberately frame and push simple descriptions of targeted environmental problems in order to mobilize broad support during the problem-definition stage. Policy entrepreneurs use triggering problems or focusing events such as perceived ecological catastrophes and environmental disasters to document the need for environmental policy, to build public support for environmental policy, and to justify government involvement.[21] Environmentalists, scientists, and mass media are important in raising public awareness and increasing the visibility of particular environmental issues, but also politicians actively focus the general public's attention on environmental problems and mobilize public and political support for environmental policy. Because of their tremendous impact on problem definition, transnational coalitions of policy entrepreneurs spend time, energy, talent, and reputation on spreading ideas and advocating policy solutions.

Transnational coalitions of policy entrepreneurs persuade policy makers to adopt domestic environmental policies and to build environmental regimes to achieve cooperation among countries. Policy makers and the general public are most likely to respond to those international and global environmental problems that have obvious domestic effects and consequences. Individuals in transnational entrepreneur coalitions who hold high-level positions in government and public administration advocate a "go first" or "lead country" strategy indicating to other countries the seriousness of targeted environmental problems, demonstrating ability and willingness to deal with them, and highlighting the need for cooperation. Those countries that policy entrepreneurs persuade to go first play a leadership role in the regime-formation process by advocating cooperation to other countries. Under these circumstances, policy makers' primary motivation is protection of the environment, not protection of national economic interests or expansion of political power.

Individuals in transnational coalitions of policy entrepreneurs who are high-level officials of international organizations play crucial leadership roles at several stages in regime formation. As prominent representatives of international organizations, they can raise global awareness of targeted environmental problems and build public and political support over the entire regime-formation process. In the problem-definition and agenda-formation stages, they initiate processes of persuasion and discovery of common interests and values, and they assist states in identifying joint gains.[22] They can also initiate and manage processes in which a broad range of nonstate actors, including prominent individual scientists and ecologists, specialized organizations of the United Nations, and international mass media, frame global environmental problems and place them on the international environmental agenda. In the pre-negotiation phase, they can establish important negotiation arenas. In order to rapidly and successfully complete the negotiation phase, policy entrepreneurs placed in international organizations attempt to increase the flexibility and the distributive fairness of environmental regimes. And, by focusing public expectations on international environmental negotiations, they help to secure signatures of states at a later stage.

I also present a detailed empirical study of the recent dramatic change of the global ocean dumping regime. In 1983 an anti-dumping transnational coalition imposed a moratorium on radwaste disposal, and in 1993 agreement was reached on a permanent global radwaste disposal ban. This regime change can be explained satisfactorily only if the actions of an ENGO are taken into account, independent of state interests, epistemic communities, or other more privileged explanatory approaches. An ENGO acted as a catalyst in the transformation of the regime and was instrumental in changing the regime's norms, rules, and principles. Beginning in the late 1970s, Greenpeace International, a well-staffed, professional, global ENGO, mobilized national and international public opinion and strengthened a transnational anti-dumping coalition by mobilizing other ENGOs, special-interest groups, and mass constituencies. Moreover, Greenpeace attacked the scientific and regulatory principles and norms of the regime, and monitored compliance by states. Which are the principal propositions regarding the influence of ENGOs emanating from this study? The study suggests that ENGOs influence environmental regimes primarily by mobilizing international public opinion, by building transnational environmental coalitions, by monitoring the environmental commitments of states, and by advocating precautionary action and environmental protection.

This case documents that even powerful states can be pressured to adjust their environmental policy in accordance with regime norms and principles. ENGO pressure was mediated and enhanced by the global ocean dumping regime. Three mediating effects of the regime should be noted. By serving as a global institutional focal point for the governmental and nongovernmental opposition to radwaste disposal, the regime strengthened the anti-dumping coalition. Also, within the framework of the regime, the anti-dumping governmental coalition adopted resolutions and treaty amendments aimed at halting radwaste disposal. This significant legal transformation affected the regime's principles, norms, and rules. Finally, the anti-dumping governmental coalition used the regime to establish global behavioral norms and standards that put pressure on countries to make their policies more environmentally acceptable. The regime's norms and standards significantly raised the political costs of noncompliance.

To sum up: I show in this book that, under certain circumstances, transnational policy entrepreneurship and public ideas are important in the development of environmental regimes. In particular, transnational coalitions of policy entrepreneurs and ideas are crucial in establishing the underpinning norms and principles of environmental regimes. To build political support, to establish legitimacy, and to mobilize others, policy entrepreneurs engage in persuasion, debate, campaigning, and public diplomacy—that is, they mobilize public opinion as a sanction on stubborn negotiators and governments.[23] Under some circumstances, the norms and principles of a global environmental regime are constructed and changed in response to powerful public ideas that identify a common environmental problem, its causes, whose responsibility it is, and how it is to be solved. As powerful public ideas influence how states define their interests, objectives, and values, they also define the intellectual contents of the norms and principles that underpin environmental regimes.

Global environmental regimes are neither permanent nor static; they develop and change over time. Regime formation is best understood as a sequence of stages and processes. Regime change also unfolds in stages and processes, and important similarities exist between regime formation and regime change. First, public ideas and international public opinion sometimes create the deep normative structure of environmental regimes. Second, by shaping, spreading, and consolidating public ideas, policy entrepreneurs (such as prominent scientists and ecologists, environmentalists, political leaders, and national and international government officials) establish international policies and institutions for environmental protection. Third, scientific experts' influence over the development of an environmental regime is dramatically reduced, irrespective of consensus among scientists, when their advice collides with a powerful public idea. By defining the deep normative structure of environmental regimes, widely accepted principled ideas about environmental problems are often more influential than scientists' causal ideas about environmental problems.

Methodology

It was the international trend toward banning ocean dumping of low-level radioactive waste, which became prominent in the early 1980s, that first caught my attention. The issue of radwaste disposal certainly seemed politically sensitive as well as institutionally significant. In the words of the head of the regime secretariat and a former London Convention conference chairman, this issue was of "extreme importance to the operation of the [London Dumping Convention]" (Nauke and Holland 1992, p. 75). I wished to know how different countries looked upon radwaste disposal, the knowledge and advice of scientific experts, and the views of ENGOs regarding this issue. I soon learned that radwaste disposal was regulated under the global ocean dumping regime, created in the early 1970s. Hence, to understand the developments in the 1980s and the 1990s, and the underpinning regime norms and principles, as well as the regime rules and decision-making procedures, required that an additional set of questions be answered: What had caused the creation of the global ocean dumping regime in the first place? How had this environmental regime come into existence? Why was radwaste disposal regulated under this regime?

As I have already mentioned, ocean dumping of radioactive waste began in 1946. However, before the creation of the global ocean dumping regime (1972) there existed no regime for controlling ocean dumping of wastes, including radioactive ones. During the early stages of my research, I therefore examined whether scientific evidence that ocean dumping was seriously threatening the oceans of the world had persuaded countries to build the regime. Another possible explanation was that public pressure had motivated politicians to act. Alternatively, perhaps countries did not consider ocean dumping a significant environmental issue but were primarily concerned with increasing political power or protecting economic interests. Careful empirical research was indeed needed to shed light on these questions.

Environmental issues are less straightforward than most of the political, economic, and military security issues that states deal with. They are often both science-intensive and trans-scientific. Hence, most environmental issues have a clear ethical and normative dimension. It therefore is necessary to examine how states look upon environmental issues, and how and why their perception may change. For instance, why is a problem that not long ago was considered insignificant, or perhaps even was ignored, suddenly seen as an important international environmental issue? What explains why an issue, after some time, is looked upon in a new and different way? Under certain conditions, technological development, scientific discovery, and altered distribution of capabilities among states influence regimes. But ideas, actors, and the uptake of ideas change too, and it is necessary to carefully examine the ways in which ideas and actors interact to influence regimes. Surprisingly little has been established about the impact of policy entrepreneurs and public ideas on global environmental regimes. Detailed empirical studies indeed seem useful.

Because most social decisions are multicausal, it is necessary to trace the different influences, show how they have impact, and measure and compare them in relation to each other. I have followed the research strategy recommended by Levy et al., who emphasize "the value of careful efforts to reconstruct the creation stories of individual regimes through procedures like process tracing and thick description" (1995, p. 287). "The challenge before us in improving our understanding of regime formation," Levy et al. continue, "is to delve into the subtleties of these interaction effects as they unfold in individual cases, without losing track of the importance of identifying patterns that can sustain useful generalizations about the creation of international regimes."[24] To date, a limited number of global environmental regime-formation processes have been examined in detail. Careful case studies conducted within a coherent theoretical framework are needed in order to further develop and compare theoretical hypotheses and models. A single case study is particularly useful for the purposes of comparing propositions of established theories and developing a new line of argument that is supported by the case.

The formation of the global ocean dumping regime has not yet been empirically documented in detail. Neither has the recent change of the regime.[25] This book is based on a broad range of empirical data and information sources: original source documents; documents obtained from the archives of Denmark's Ministry of Foreign Affairs; government reports; reports of congressional hearings in the United States; unofficial memos of members of the secretariat of the 1972 United Nations Conference on the Human Environment; articles from newspapers, magazines, and professional journals; interviews and correspondence with participants and close observers; published accounts of participants; and my own observations. Many relevant documents exist in the public domain (for example, those produced by the London Convention secretariat at the International Maritime Organization in London), and my empirical research benefited greatly from their availability. Several interviewees read and commented on parts of draft manuscripts; they considerably improved my understanding of the global ocean dumping regime and the radwaste disposal issue.

Is global regulation of ocean dumping of radwaste a rare or unusual case, or can lessons based on it be generalized to other cases? Several recent studies have demonstrated that transnational coalitions spreading normative ideas and powerful images can be very influential in international politics. For instance, by engaging in moral persuasion and mobilizing social pressure, a transnational NGO coalition recently succeeded in establishing an international norm banning antipersonnel land mines (Price 1998). Another security study has examined the international taboo on the use of chemical weapons and has shown that socially constructed symbols, rather than objective facts and technical data, can have important political consequences for warfare (Price 1995).[26] And "transnational moral entrepreneurs" mobilizing popular opinion and public support have played a critical role in the creation of global norms prohibiting piracy, slavery and the slave trade, international drug trafficking, and the killing of whales and elephants (Nadelmann 1990). It seems evident, therefore, that the propositions based on this single case study are relevant to several other international issue areas.

It is worth stressing that the substantive areas examined in those studies range from the environment to military security—in other words, from "low politics" to "high politics." Despite some minor differences, these cases exhibit a number of important common features that also are central themes of this book: Cooperation aims at protection or provision of an international public good. Transnational coalitions are significant actors. International organizations might be useful policy arenas but are only seldom important actors. The issue of conflict becomes salient at crucial points in the international policy process. Campaigning, persuasion, and public diplomacy are important instruments of influence and control. Policy entrepreneurs use ideas, images, and symbols strategically to focus public and policy attention on the targeted issue.

Moreover, these common features tend to manifest themselves in a number of characteristic ways: Even powerful states often need to reformulate their policies and sometimes their interests. These issues pit civil society against the state. Norms and ideology carry equal and often more political weight than scientific knowledge and technical facts. International ideas and norms reflect Western ideas and norms.

Outline of the Book

In chapter 2 I give an overview of the history of ocean dumping of low-level radioactive waste. In chapter 3 I develop and discuss the concept of transnational entrepreneur coalitions and place it in a broader empirical and theoretical framework; the power-based, interest-based, and knowledge-based analyses of regimes and cooperation are also examined and compared. In chapters 4–6 I present empirical evidence bearing on the three prominent regime-analytical approaches. In chapter 4 I focus on the scientific advice given to the U.S. Congress, the Nixon administration's approach to ocean dumping, and the involvement of international scientific organizations in the ocean dumping issue before the 1972 United Nations Conference on the Human Environment (the so-called Stockholm conference). In chapter 5 I document the domestic policy process in the United States, the tremendous political significance of the "dying oceans" idea, and the formulation of U.S. foreign policy on ocean dumping. In chapter 6 I examine the negotiations on the global ocean dumping regime, including the role played by policy entrepreneurs inside the United Nations. Taken together, chapters 4–6 support the conclusion that regime formation did not happen as predicted by the power-based, interest-based, or knowledge-based approaches to regime analysis.

In chapter 7 I present empirical evidence that further documents the interplay between a transnational coalition of policy entrepreneurs and the "dying oceans" idea in this process of regime formation. The strengths and weaknesses of regime analysis that is primarily concerned with power in a material sense, with egoistic self-interest, and with scientific-technical knowledge are also examined in the light of the case. I conclude that the formation of this regime can be explained adequately only if one follows an ideational approach stressing the impact of ideas and transnational entrepreneur coalitions on international environmental politics.

In chapters 8 and 9 I examine the more recent regime development on radwaste disposal. I document and analyze a significant change of this regime; the topic of regime change has so far remained relatively underexplored in regime analysis.[27] In chapter 8 I document the development from the late 1970s until the 1993 decision to globally ban radwaste disposal, showing that an ENGO acted as a catalyst for change in the global ocean dumping regime and forced the transformation from regulation to prohibition even though scientists and powerful states supported radwaste disposal. In chapter 9 I suggest four general propositions about the influence of ENGOs on environmental regimes. I also discuss the strengths and weaknesses of the three prominent approaches to regime analysis in the light of the 1993 global radwaste disposal ban. In chapter 10 I point out the limitations of this case study, discuss the generalizability of the findings, and suggest how further research might shed more light on the issues raised earlier in the book. Finally, I discuss whether the decision to ban radwaste disposal should be considered "good" international environmental policy.

2
History of Efforts to Control Ocean Disposal of Low-Level Radioactive Waste

Radioactive Wastes—the Source of the Problem

Experts have recognized since the early 1920s that radioactive materials can have detrimental effects on human beings and on the environment (Lapp 1979). Even low-level radioactive waste, the least hazardous of radioactive materials, must therefore be kept isolated.

Radiation is a process in which unstable atomic nuclei spontaneously disintegrate, usually emitting ionizing radiation (alpha and beta particles, gamma rays, and x rays). Radioactivity is a *nuclear* phenomenon, and it is independent of any chemical or physical changes that the *atom* may undergo. Ionizing radiation is a form of radiation that is able to ionize atoms in the matter through which it passes.[1] An alpha, beta, or gamma ray entering a piece of matter transfers energy to the matter through collisions with the atoms in it. The particle can pass completely through the matter—and lose only a portion of its original energy—if the matter is sufficiently thin or if the radiation has a high energy. If not, the particle will be absorbed in the matter and will lose its energy through ionization. Ionization causes chemical reactions and general heating of the matter.

Alpha particles are capable of traveling only a few inches in air and are easily stopped by a sheet of paper or by intact skin. They can, however, cause great damage if they are ingested or inhaled. Alpha particles produce more deleterious biological effects than the lightly ionizing radiation associated with beta, gamma, or x radiation. Beta particles are smaller and faster than alpha rays and have a greater range. Beta particles can penetrate several layers of tissue and can cause burns when skin is exposed, but a relatively thin layer of water, glass, or metal can stop them. A beta emitter (a radioactive substance that emits beta particles) presents a greater hazard if it enters the body, though generally less of a hazard than an alpha emitter. Gamma rays, a very high-energy form of radiation, penetrate a relatively great thickness of matter, such as steel and concrete, before being absorbed. Overexposure of the body to gamma radiation results in deep-seated organic damage. Gamma radiation is by far the most serious external hazard of the three types of radiation from radioactive substances, and it requires heavy shielding and remotely controlled operations.

One curie (Ci) is defined as the amount of radioactive material that will produce 37 billion (3.7×10^{10}) disintegrations per second. This is approximately the number of disintegrations per second in a gram of radium. A more recent unit is the becquerel (Bq), which is the amount of radioactive material that produces one disintegration per second. One terabecquerel (TBq) is approximately equal to 27 curies. Minute radiation is measured in nanocuries, one nanocurie being a billionth (10^{-9}) of a curie.

The two basic options for managing radioactive wastes are (1) to concentrate and then store them and (2) to dilute them and then disperse them in the environment.[2] Many forms of radioactive wastes exist, and all must ultimately be disposed of. *Disposal* signifies the final fate of the waste without the intention of retrieval later; *storage* means a temporary scheme.[3] Disposal does not imply continuous monitoring; storage does. Low-level radioactive waste, which also can be disposed of on land, is the only form of radioactive waste that has been dumped in the oceans.[4] Most of this waste has been packaged in 55-gallon drums filled with concrete to ensure that the drums would sink to the ocean bottom. These drums were not designed to remain intact for long periods on the sea bottom, and it was assumed that the contents would be released almost instantly (Templeton 1982, p. 39).

Fallout from atmospheric testing of nuclear weapons, which also has reached the oceans, is regulated under the Limited Test Ban Treaty.[5] Volcanic eruptions on the sea bed, fallout from atmospheric nuclear weapons tests, the Chernobyl nuclear disaster, and operational discharges from the British Sellafield nuclear reprocessing plant (previously known as Windscale) have been by far the greatest contributors to the radioactivity of the marine environment. As the marine scientists Bewers and Garrett summarize (1987, p. 106): "The total amount of radioactivity dumped in the ocean, some 6×10^4 TBq, is much less than the approximately 2×10^8 TBq that were added to the oceans as a result of atmospheric testing of nuclear weapons between 1954 and 1962. This, in turn, is only 1% of the 2×10^{10} TBq that exists naturally in the ocean."[6] Nuclear accidents—the most important being those at Windscale (1957), Three Mile Island (1979), and Chernobyl (1986)—have contributed insignificantly to "the ocean inventory of radionuclides on a global scale" (GESAMP 1990, p. 51).

There are many definitions of radioactive waste, including those of the International Atomic Energy Agency (which assists the global ocean dumping regime with definitions of radioactive waste) and those of individual laboratories. The most important forms of waste are *high-level waste*, *transuranic waste*, *mill tailings*, and *low-level waste*. These categories of waste are not simply a function of the nature of the waste. Scientists and regulators have developed subjective distinctions.[7]

There are basically two sources of high-level waste: unreprocessed spent fuel assemblies from nuclear power plants and the highly radioactive waste from reprocessing plants. The waste contains the fission products and actinides (heavy elements) separated from the dissolved fuel. The highly toxic nature of high-level waste, as well as the extremely long half-lives of the nuclides contained in such waste, requires that it be isolated from the biosphere for several thousand years.[8] In the United States, most high-level waste is currently kept in pools or dry casks at reactor sites while a site for a mined geological repository is sought.[9] Emplacement of high-level waste into the seabed was debated and rejected at the 1983 annual meeting of the signatories to the global ocean dumping convention.[10]

Transuranic waste, which results primarily from spent-fuel reprocessing and nuclear weapons production, contains transuranic elements (that is, elements having a greater atomic number than uranium in the periodic table—for example, plutonium, americium, and neptunium) in concentrations greater than 10 nanocuries per gram (Green and Zell 1982, p. 115). The most prominent element in most waste of this type is plutonium. The risks from transuranic waste are not a function of the actual concentration of radioactive contaminant but rather lie in the fact that these elements decay so slowly as to remain radioactive for an extremely long time.

Mill tailings are, for the most part, produced at uranium mills. For every ton of uranium ore that is milled in the United States, not more than about 2 kilograms of uranium is extracted. The rest of the ore is discharged as fine-grained, sand-like, and silty materials. Mill tailings contain relatively low concentrations of other naturally occurring radioactive materials, the most important being radium (which decays to produce radon). Huge piles of tailings have accumulated at numerous sites. Typically, tailings are left near the uranium mills where the ore is processed. Today, regulators see the accumulation of huge piles of tailings as an unacceptable waste disposal practice. The best solution is to require burial of the tailings well below grade, or below the surface of the adjacent terrain.[11] So far, mill tailings have not been a subject of long and severe public protests.

In the United States, low-level waste is defined by the Low Level Radioactive Waste Policy Act as radioactive waste not classified as high-level radioactive waste, transuranic waste, spent nuclear fuel, or by-product materials as defined in Section 11e(2) of the Atomic Energy Act (uranium or thorium tailings and wastes).[12] It is, in other words, waste that does not fall into any of the other three categories. Low-level waste can take solid, liquid, or gaseous form. Typical waste includes contaminated protective shoe covers and clothing, wiping rags, mops, filters, reactor water treatment residues, equipment and tools, luminous dials, medical tubes, swabs, injection needles, syringes, and laboratory animal carcasses and tissues (U.S. Nuclear Regulatory Commission 1996, p. 11). Low-level radioactive waste is generated by hospitals; by medical, educational, or research institutions; in private or government laboratories; and at facilities forming part of the nuclear fuel cycle (e.g., nuclear power plants, fuel fabrication plants). The overall volume of low-level waste produced by commercial and government sources in the United States has increased steadily, especially since the late 1970s. Approximately 4 million cubic meters of waste were produced in 1990, and the waste volume is expected to almost double by 2020 (U.S. Environmental Protection Agency 1994, p. 2). At the same time, as a result of the dramatically rising costs of disposal, which have led to better packaging of the waste, the volume of disposed waste has decreased.

International Ocean Disposal before 1973

Beginning as early as 1946 and extending into 1972, ocean disposal of low-level radioactive waste was practiced by a number of states without international controls. Disposal operations were carried out under the direction of the relevant national authorities. Great Britain was the principal dumping country between 1949 and 1970. Isolated instances of ocean disposal were carried out by Belgium in 1960, 1962, and 1963. Table 2.1 lists the dominant countries that conducted ocean disposal operations, the number of containers dumped, and the associated radioactivity.[13]

International efforts to control dumping of radioactive materials date back to the 1958 United Nations Conference on the Law of the Sea (UNCLOS), which adopted an article stating that "every State shall take measures to prevent pollution of the seas from the dumping of radioactive waste, taking into account any standards and regulations which may be formulated by the competent international organizations."[14] It also said that "all states shall co-operate with competent international organizations in taking measures for the prevention of pollution of the seas or air space above, resulting from any activities with radioactive materials or other harmful agents."[15] This article—a compromise between states that engaged in such practices and other states (especially the Soviet Union) that favored a complete prohibition of nuclear waste dumping—had no great effect

Table 2.1
Countries engaged in unilateral ocean dumping of radioactive waste. Radioactivity is at the time of dumping. Source: Holcomb 1982, p. 184. The data on dumpings by the Republic of Korea are from LDC 1985a, Annex 2, p. 113. The more accurate data on U.S. containers and approximate radioactivity (4.3×10^{15} Bq), but not weight, are from Hagen 1983, p. 49.

			Approximate weight (metric tons)	Number of containers	Approximate radioactivity (Ci)
U.S.	1946–1967	Pacific, Atlantic	25,000	107,000	116,100
U.K.	1949–1966	Atlantic	47,664	117,544	143,200
	1968, 1970	Atlantic			
Netherlands	1965–1972	Atlantic	935	2,365	62
Japan	1955–1969	Pacific	656	1,661	452
Rep. of Korea	1968–1972	Sea of Japan		115	NA*
			74,255	228,685	259,814

* not available

(McDougal and Burke 1985, pp. 864–868). Although states pledged to cooperate and take relevant "measures," neither the precise nature of these measures nor any minimum standards were specified (Schenker 1973, p. 37). "In essence," writes Frye (1962, p. 29), "the conference produced no community policy at all on the matter."[16]

The conference resolution, which did not have the force of a treaty, further recommended that the International Atomic Energy Agency, an agency of the United Nations that promotes the peaceful use of nuclear power, undertake studies of the technical and scientific problems connected with sea disposal of radioactive waste. An IAEA panel established for this purpose concluded in 1961 that, although there was no general hazard at that time associated with ocean dumping of radioactive waste, such hazards could become significant in the future.[17] Further, the panel concluded that ocean disposal of high-level waste could not be recommended and that low-level waste should be dumped only under controlled conditions. The panel proposed an international accord to keep radioactive hazards from accumulating to unacceptable levels. However, this proposal was never embodied in an international convention.[18]

Starting in 1967, a voluntary mechanism set up by the Nuclear Energy Agency of the Organization for Economic Cooperation and Development provided guidelines and undertook supervisory responsibility for disposal of low-level waste from NEA member countries (Holcomb 1982, p. 184). That same year, the first NEA-supervised international nuclear waste dumping operation was carried out, at a depth of 5000 meters in the eastern Atlantic. The primary objectives of the NEA were "to develop, at the international level, a safe and economic method for ocean disposal and to demonstrate this by a joint experimental disposal operation involving several member countries."[19] Belgium, France, the Federal Republic of Germany, the Netherlands, and Britain supplied some 35,000 containers of waste weighing nearly 11,000 tons and containing approximately 8000 Ci of radioactivity. The primary dumpers—Belgium, the Netherlands, Switzerland, and Britain—participated in a series of coordinated dumping operations that took place in 1967, in 1969, and in each year from 1971 to 1982. France, Italy, Sweden, and the Federal Republic of Germany participated in only the first two dumping operations. The regionally coordinated dumping called for agreement on dump-site selection, package design for the waste material, facilities available on the dumping vessel, and duties of escorting officers.[20] The principal sources were nuclear power plant operations; other nuclear fuel cycle operations (including fuel fabrication and reprocessing); radionuclides used in medicine, research, and industry; and the decontamination and dismantling of redundant plant and equipment.[21] Table 2.2 lists the mass of the material dumped from 1967 to 1982 as well as the estimated alpha and beta/gamma activity of the waste at the time of packaging.

From 1946 through 1970, the U.S. Atomic Energy Commission permitted the disposal of low-level radioactive waste in the ocean at AEC-licensed sites. Approximately 107,000 canisters (116,100 Ci) were disposed of.[22] The waste consisted of contaminated laboratory glassware, bench tops, floor coverings, tools, chemicals, and animal carcasses. The waste was

Table 2.2
Radioactive waste dumped into North Atlantic under supervision of Nuclear Energy Agency, 1967–1982. Source: LDC 1985a, Annex 2, p. 104.

	Gross weight (metric tons)	Approximate radioactivity (Ci)		
		Alpha	Beta/gamma	Tritium
1967	10,895	253	7,636*	
1969	9,178	485	22,066*	
1971	3,968	627	11,148*	
1972	4,131	681	21,626*	
1973	4,350	740	12,660*	
1974	2,265	416	100,356*	
1975	4,454	767	57,374	29,690
1976	6,772	878	53,518	20,703
1977	5,605	958	76,451	31,886
1978	8,046	1,101	79,628	36,613
1979	5,416	1,414	83,166	42,240
1980	8,319	1,855	181,227	98,135
1981	9,434	2,177	153,566	74,372
1982	11,693	1,428	126,988	77,449
1983	ND†			
1984	ND			
	94,526	13,780	987,410	411,088

* includes tritium (^{3}H)
† no dumping

mainly disposed of in three sites in the Atlantic (off New Jersey and Massachusetts) and one site in the Pacific (off San Francisco). These four sites received more than 90 percent of all the radioactive waste containers and 95 percent of the estimated radioactivity dumped.

The largest quantity of radioactive waste was dumped in the period 1946–1962. In 1960, because of increasingly strong public opposition to ocean disposal, the AEC imposed a moratorium on the issuance of new licenses for dumping (Mazuzan and Walker 1985, pp. 344–372). The AEC turned instead to land burial, which also entailed relatively lower costs compared to ocean disposal. Contributing to the high costs were containers, transportation to the dock, and transportation to disposal point in the ocean (Straub 1964, p. 326). By 1963 most ocean dumping had been phased out. About 350 containers (with an estimated activity of 230 Ci) were dumped into the ocean between 1963 and 1970 when radwaste disposal was terminated (Holcomb 1982, p. 189).[23]

International Ocean Disposal after 1973

Representatives from 92 states, meeting in a highly publicized UN-sponsored conference held in London from October 30 to November 13, 1972, agreed for the first time to establish a global environmental regime controlling the disposal of wastes in the oceans, radioactive wastes included. All the Western European maritime and nonmaritime states participated, as did the Soviet Union, the United States, Canada, Japan, Australia, and New Zealand. A large number of developing countries were also represented.[24] This truly global environmental regime prohibits ocean dumping of high-level radioactive waste while allowing medium-level and low-level radioactive waste to be dumped when the dumping is done in essentially a controlled way.

The global ocean dumping regime prohibits dumping without a permit. Governments are responsible for issuing permits to dumpers under their jurisdiction and for seeing that any required conditions are fulfilled. Members must report the quantity and the nature of the material dumped to a secretariat, which then reports this information to the other members of the regime. The successful implementation of the London Convention depends on the development of an effective enforcement mechanism by each regime member; enforcement is the responsibility of each signatory. However, it is the global dumping regime that determines the criteria for issuing radwaste dumping permits—in essence the regime's regulatory policy on radioactive waste disposals—and dumping criteria are regularly reviewed by the members. This takes place at the London headquarters of the International Maritime Organization, a United Nations agency that facilitates international cooperation on technical matters affecting international shipping and serves as the regime's secretariat. The International Atomic Energy Agency determines what radioactive materials are unsuitable for ocean dumping and makes recommendations on the disposal of other radioactive wastes. The IAEA also makes recommendations with regard to selection of dumping sites, packaging for dumping, approval of the ship and its equipment, escorting officers, and record keeping. In setting radiation protection standards, the IAEA relies on the recommendations of the International Commission on Radiological Protection, an international nongovernmental scientific organization of professional radiologists.

After 1971, only Belgium, the Netherlands, Switzerland, and Britain had annually conducted radwaste disposal in the eastern Atlantic.[25] As I have mentioned, France, Italy, Sweden, and the Federal Republic of Germany withdrew from the operation of the NEA arrangement in 1974, opting for land storage of waste products.

In 1979, however, Japan and the United States announced intentions to initiate new programs of radioactive waste dumping into the ocean. Japan became a member of the global dumping regime in 1980, most probably in order to gain legitimacy for its plans to dump as many as 2 million drums, containing about 100,000 Ci of low-level waste per year, into the Pacific.[26] The United States considered a plan to scuttle aging nuclear submarines in the Atlantic and the Pacific. As many as 100 submarines were to be scuttled, each representing 50,000 Ci of radioactive waste. The U.S. Environmental Protection Agency formulated new regulations that would permit resumption of ocean dumping of radioactive waste, including decommissioned nuclear submarines (Norman 1982, pp. 1217–1219). But Congress and environmentalists soon defeated these plans, and the radioactive engine compartments of the submarines were instead buried at two government land facilities as an "interim move" (Trupp 1984, pp. 34–35).

In 1983, in response to the planned dumping by Japan and the United States and to the annual European dumping operation,[27] a transnational coalition led by ENGOs,[28] developing countries,[29] and the Nordic countries in an alliance with Spain succeeded in imposing a moratorium on radwaste dumping within the global ocean dumping regime. Although the moratorium did not technically outlaw dumping, the previous practices were no longer tolerated internationally. The moratorium was strongly opposed by the pro-dumping states, particularly Britain and the United States. The British government threatened to renounce its membership of the global dumping regime with a view to further dumping, and the Swiss government announced that it would continue dumping low-level waste in spite of the moratorium. France (which had last dumped in 1969) and the Netherlands also seemed interested in further dumping.[30] But leading British transport unions boycotted the planned dumping by Britain, and the government shelved its plans to dump. In December of 1984, responding to protests from its Pacific neighbors, the Japanese government decided to cancel its ocean dumping program.[31]

Significantly, however, the members of the global ocean dumping regime continued to meet and avoided undermining the regime's authority. This occurred despite conflicts surrounding seabed emplacement of high-level radioactive waste, and the 1984 annual meeting agreed that the regime was "the appropriate international forum to address the question."[32] In a similar manner, the group of international experts that met in 1985 within the context of the regime to examine whether radwaste disposal was advisable on technical and scientific grounds concluded that "present and any future dumping can only take place within the still-developing framework of international regulations."[33]

Contrary to the interests of the pro-dumping states, the transnational anti-dumping coalition renewed the radwaste disposal moratorium for an indefinite period in 1985. A decision whether to continue or discontinue radwaste disposal would not be made until environmental aspects as well as wider political, legal, economic, and social aspects of radwaste dumping were examined. Subsequently, the members of the global dumping regime tried without success to reach agreement on the environmental effects of radwaste disposal. The United States, Britain, and Japan, occasionally supported by France, regularly opposed the waste management policy advocated by the majority of nations and refused to incorporate global policy on this issue into their national policies.[34] A political deadlock appeared inevitable.

Since the mid 1980s, discharges of radioactivity from Sellafield into the Irish Sea had also caused concern within the global ocean dumping regime.[35] In the spirit of this regime, and also in the spirit of a regional dumping regime composed of countries bordering the North Sea (a regime established by the so-called Oslo Convention), the Sellafield discharges were no longer a matter involving only Britain and Ireland. Characteristically, political parties and Greenpeace Denmark put pressure on the Danish government, which also was concerned about its fishing interests, to protest against the Sellafield discharges.[36]

Beginning in the 1980s, American, British, and Japanese attempts to find permanent land-based disposal facilities for radioactive waste were increasingly met with public and political opposition, and acceptable solutions seemed out of reach. To a large degree, they all lacked sufficient land-based waste disposal facilities for low-level radioactive waste (see table 2.3). No new waste disposal facilities had been opened in the United States since 1971, and, accordingly, disposal at sea was constantly being reexamined. In the late 1980s, U.S. EPA officials were reconsidering radioactive waste disposal in the oceans, but dumping was not resumed.[37] Similarly, the British Ministry of Defense considered dumping obsolete nuclear submarines in the ocean, in possible breach of both the United Nations Convention on the Law of the Sea (which was signed in 1982 and took effect in 1994) and the moratorium decided upon by the global ocean dumping regime.[38]

In 1993 the international review of radwaste disposal was completed. That same year, the United States and Japan reversed their positions on this issue, and agreement was reached within the regime on a global ban on radwaste disposal. Although strongly opposed to the ban, Britain and France decided to accept it. The radwaste disposal ban marks a significant regime change from a permissive allowance of ocean dumping of low-level radioactive waste to a new emphasis on precaution and prevention.[39]

Summing up a complex regime development, the London Convention has established a robust global environmental regime for regulation of ocean disposal of low-level radioactive waste. Significant institutional changes have "transpired according to an *ex ante* plan (and hence part of

Table 2.3
Status of low-level and intermediate-level waste disposal facilities in Britain, France, Japan, and the United States, 1993. Source: IAEA 1994, pp. 93–94.

	Site name	Repository concept
Site selected		
U.S.	Wake County, North Carolina	Engineered near-surface facility
Under licensing		
U.S.	Ward Valley, California	Engineered near-surface facility
U.S.	Boyd County, Nebraska	Engineered near-surface facility
In operation		
France	Centre de l'Aube (1992)*	Engineered near-surface facility
Japan	Rokkasho (1992)	Engineered near-surface facility
U.K.	Drigg (1959)	Simple and engineered near-surface facilities
U.S.†	Hanford, Washington (1965)	Engineered near-surface facility
Nearing closure		
France	Centre de la Manche (1969)	Engineered near-surface facility
U.S.	Barnwell, South Carolina (1971)	Simple near-surface facility

* date operation started
† A disposal facility at Beatty, Nevada, which had been in operation since 1962, was closed on January 1, 1993. See U.S. General Accounting Office 1995, p. 12.

the original institution) for institutional change" (Kenneth A. Shepsle, quoted in Ostrom 1990, p. 58).[40] This book will document that a transnational coalition of policy entrepreneurs provided crucial leadership in the regime-building process. In the early 1970s, a transnational coalition of prominent scientists and ecologists, politicians, and high-level United Nations officials mobilized international public opinion, raised international expectations, and succeeded in building a global environmental regime based on a U.S. initiative. When radwaste disposal was banned in 1993 as the culmination of a process initiated in the late 1970s, a global environmental pressure group played an important catalytic role by mobilizing and focusing international public opinion on radwaste disposal, by building an influential transnational environmental coalition, by monitoring activities regulated under the regime, and by advocating a precautionary approach to environmental protection. Transnational actors have created global norms and regulatory machinery constraining pro-dumping governments.

Summary

In 1972, states reached the first agreement on a global environmental regime regulating ocean disposal of low-level radioactive waste and prohibiting the dumping of high-level radioactive waste into the world's oceans. This global environmental regime has functioned as intended. It has been an active international forum for policy coordination, and a significant policy shift—the 1993 radwaste disposal ban—has been adopted. International environmental agreements and conventions often fall short of expectations.[41] Nevertheless, international policy coordination and cooperation has taken place in this issue area. Thus, the London Convention is "widely regarded as one of the more successful regulatory treaties" (Birnie and Boyle 1992, p. 321).

Membership in the global dumping regime has been rising since 1972. By 1984, 56 governments had become members of the regime.[42] By 2000, the number had increased to 78.[43] Governments that had been involved in radwaste disposal as well as governments opposing this practice have become members of this environmental regime. It thus represents a significant departure from individual states' uncontrolled dumping in the past. Today the global dumping regime is the forum where states debate their dumping policies and attempt to reach agreement on international controls. And although a considerable number of regimes exist in the field of marine pollution prevention, this regime is generally considered one of the most prominent in its field. One U.S. EPA official said in 1988: "Although in the past decade the London Convention has been supplemented by several regional marine protection agreements to address the particular needs of such areas as the Mediterranean, the Caribbean, and the South Pacific, it is still generally recognized as the standard against which all disposal operations must be measured." (Sielen 1988, pp. 4–5)

Until 1972, marine environmental concern had been limited to oil pollution. Unlike the London Convention, earlier conventions on nuclear ships and nuclear damages treated certain activities as dangerous, not as specifically environmental problems.[44] Another noteworthy and important characteristic is that the global ocean dumping regime regulates intentional, as opposed to accidental and unintentional, disposal of waste into the ocean. Thus, the regime considers ocean dumping a conscious decision to change a given environmental medium for perceived social benefits.

In a world searching for policy responses to global environmental problems such as global warming and ozone depletion, the global ocean dumping regime—one of the first global environmental institutions addressing problems of marine pollution—is a significant case. Our knowledge of global cooperation on environmental protection is still limited, and only a few studies have examined how global environmental regimes are built and changed. This case offers a window on the global politics of regime formation and regime dynamics. It can therefore offer theoretical as well as practical insight into global cooperation on environmental protection.

3
Transnational Coalitions of Policy Entrepreneurs, Regime Analysis, and Environmental Regimes

In this chapter I present the *transnational entrepreneur coalitions* (TEC) approach. I also explore the power-based, interest-based, and knowledge-based approaches to the study of regimes and cooperation. In particular, I examine how these three types of explanation envision the construction of the global ocean dumping regime and how and why this regime, once constructed, would change. In this way, I conceptually prepare the ground for comparing the TEC approach and the three regime approaches.

The TEC approach is concerned with agency and process, not with international structure, system, and order.[1] It combines ideational analysis of policy and institutional processes with a focus on transnational ideas-based actors. The approach is dynamic rather than static, and it focuses attention on the ideational aspects of policy—i.e., the processes and actors influencing how states perceive their objectives and interests and how these change over time. Interests are not exogenously given, and they are not permanent and static in ideational political analysis. To analyze the transnational dimensions of policy entrepreneurship, the approach builds upon one variant of ideas-based analysis developed in the American literature on policy entrepreneurs and in the literature on regulatory policy making in the European Union.[2] It is a middle-range, context-specific set of propositions concerning regime processes and regime dynamics.[3] In contrast, power-based and interest-based regime models predominantly analyze regimes at the international systemic level, while knowledge-based regime analysis is primarily concerned with subnational and transnational groups. My approach is especially concerned with initiation and change of policy, with brokering, and with mobilization.

Most students of international politics subscribe to a realist point of view.[4] Realist scholars point to Thucydides's *History of the Peloponnesian*

War, Machiavelli's *The Prince*, and Hobbes's *Leviathan* as their theoretical ancestors. They share the view that states fundamentally struggle for physical survival and political independence. Realists emphasize anarchy, states as the principal actors, and pursuit of power as the primary objective of states. Realists have contributed substantially to our understanding of regimes, cooperation, and international leadership. Like many students of international politics, realists view the prospects for cooperation on environmental protection with pessimism.

Interest-based or neoliberal hypotheses stress the importance of egoistic self-interest and rationality in regime formation.[5] Although they accept basic assumptions of realism, neoliberals claim that states at times succeed in creating regimes that remedy and prevent the detrimental effects of their uncoordinated behavior. According to interest-based propositions, regimes regularize interactions among states, and under certain conditions regimes can overcome collective action problems that hinder cooperation among states. Neoliberals emphasize the importance of integrative bargaining and leadership by individuals. Comparison of interest-based and power-based approaches is possible in this case because prominent individuals and specialized United Nations agencies were closely involved in the construction of the global ocean dumping regime.

Peter Haas's *Saving the Mediterranean*, a study of the Mediterranean Action Plan ("Med Plan") to protect against marine pollution in the Mediterranean basin, has drawn attention to the question of how an epistemic community—"a network of professionals with recognized expertise and competence in a particular domain and an authoritative claim to policy-relevant knowledge within that domain or issue-area" (Haas 1992a, p. 3)—might induce states to cooperate. As already mentioned, epistemic-community theorists primarily focus on how scientists and technical experts influence international politics. Their research program has initially focused on compliance with regimes rather than regime creation. The concept of epistemic community originated in the reflective literature that asserts that cooperation fundamentally varies with the evolution and change in governments' perception of cooperation.[6] Epistemic-community analysts view the prospects of cooperation on environmental protection with moderate optimism.[7] Haas's conclusion that the Med Plan was instituted and advanced by an epistemic community radically challenges both power-based and interest-based regime analysis. Because the global ocean dumping regime is both historically and substantively comparable to the Med Plan, this study also offers a good opportunity to examine core assumptions in the epistemic-community literature.[8]

Ideas and Transnational Coalitions of Policy Entrepreneurs

This approach is concerned with the impact of transnational coalitions of policy entrepreneurs and ideas on the objectives and interests of states and, by implication, on international cooperation. The approach examines the conditions under which state behavior changes as a result of widely shared beliefs and values. The knowledge-based regime analysis is concerned with relationships between knowledge and policy outcome; the TEC approach, in contrast, explores the broader relationship between knowledge, ideas and beliefs, and policy outcome. This emerging approach was first developed in the domestic context, especially in studies on social and environmental regulation.

An important leitmotif of this approach is legitimacy as a political resource and justification of policy. Great emphasis is put on public deliberation and society's concern for its own well-being. This approach replaces atomism with institutionalism and individual self-maximization with public reasoning and problem solving. For these scholars, leadership and ideas are especially important when the public and the government develop visions about what is good for society and the nation.[9] Studies in this tradition stress that political elites and policy entrepreneurs need to win the support of broad constituencies and sometimes even of society at large in order to influence policy and define the interests and objectives of states. Though regime analysts generally have ignored such issues, John Ruggie (1998b, p. 206) has recently stressed "the role of imagery, ideas and justifications" in engaging the United States in the international order. According to Ruggie, leaders who construct state preferences are "attempting to persuade their publics and one another through reasoned discourse while learning, or not, by trial and error" (ibid., p. 202).[10] Systematic attention to public ideas and public debate about policy clearly distinguishes the TEC approach from the three prominent regime analysis approaches.

Policy entrepreneurs are individuals who significantly influence politics and policy development.[11] They can be members of Congress, civil servants, prominent experts, academics, independent advocates, or officials of international organizations. Strategically skilled, they reshape political coalitions, primarily by changing the content of policy debates. They influence policy by raising new issues or redefining existing ones, by brokering among stakeholders, and by building and broadening public and political support behind their policy solutions.

Policy entrepreneurs invest their energy, talent, reputation, and time in order to shape policy. Persuasiveness, substantial negotiating skills, and persistence are some of their most important characteristics. Sometimes policy entrepreneurs are motivated mainly by protection of private interests, such as jobs, organizational influence, and prestige. At other times, they mainly wish to promote their values and shape public policy. A particularly apt account of how policy entrepreneurs may primarily be concerned with the public, rather than private, interest can be found in a study on Jean Monnet, the intellectual architect of the European Community: "Such people are not politicians, because they hold no elected office. Yet they display great political ability. Nor can they be called bureaucrats. They may or may not operate in a formal bureaucracy. They are even anti-bureaucratic to the extent that they are impatient of routine minds. Again, though they are entrepreneurs, they are not people of business. Their imaginations are fired by the public, not a private, interest. They answer needs the citizen recognizes as his own once they have defined them. But whereas the creative politician crystallises the public's consciousness, their medium is action ahead of common awareness. They operate at the borders between politics, bureaucracy and business and belong to none. They are, for want of a better term, entrepreneurs in the public interest." (Duchêne 1994, p. 61) In such cases, increasing the overall well-being of society is policy entrepreneurs' main motivation, and they will make strong appeals to concern for the collective interest. Nonetheless, it often seems to be a combination of private interests and concern for the public interest that motivates policy entrepreneurs.

Policy entrepreneurs are most effective when a policy window opens up, and they need to act before the policy window closes again.[12] Policy windows appear when a problem demands the attention of decision makers or when decision makers are strongly motivated to solve a problem. In general, policy windows open at those relatively rare occasions when three usually separate policy streams—problems, politics, and policy ideas—come together. When policy windows appear, policy entrepreneurs push ideas and solutions as well as mobilize political and public support behind their initiative.

Individuals in transnational entrepreneur coalitions share an interest in promoting international policy projects reflecting their common beliefs and values. Transnational coalitions of policy entrepreneurs are best able to influence international policy and regimes in situations where their policy projects are compatible with powerful global public ideas and address perceived important international problems. Within particular cognitive and institutional structures, the influence of entrepreneur coalitions and the power of the ideas they push are mutually constitutive and reinforcing. These coalitions persuade countries that by creating common policies and regimes they protect both their national and their collective interests.

In the environmental field, these coalitions encompass policy entrepreneurs in government, national administrations, international organizations, environmentalists, scientists, and ecologists, who together seek to translate their shared beliefs and values into international environmental policy. Such loosely organized transnational coalitions of like-minded policy entrepreneurs held together by their shared values and common policy project might be conceived of as de facto natural coalitions.[13] Under certain conditions, their ideas and solutions become institutionalized in the form of regimes; unlike epistemic communities, however, they do not necessarily wield bureaucratic power. Their members may be found both inside and outside government, and therefore it is sometimes difficult to distinguish accurately between government and civil society. Hence, a major strength of transnational entrepreneur coalitions is their ability to transcend boundaries separating governmental structure and society and boundaries separating domestic and international arenas. They can therefore often supply policy entrepreneurship in both arenas of the two-level games through which regimes are built and changed.[14]

To solve targeted environmental problems, transnational entrepreneur coalitions strategically frame international environmental problems and deliberately construct norms and institutions defining what constitutes socially acceptable and legitimate behavior. They exert a strong influence on problem definition and agenda formation in the early stages of regime establishment. Interests in protecting the environment are often broad and diffuse, and the need for environmental protection is seldom obvious. Environmental problems do not simply "exist"; they are politically constructed. For example, Albert Hirschman (1984, p. 151) has pointed out that "some of the problems characteristic of advanced societies, such as pollution and discontent about work conditions, are more diffuse and less obviously intolerable than earlier problems of misery and oppression, so that a considerable effort of intellectual focusing and clarification is needed for these problems to enter the consciousness of the victims." Some scholars suggest using the term "issueness"—understood as "the transformation of a fact of life into a political issue"—as a way of conceptualizing the essence of the problem-definition stage (Enloe 1975, p. 11). For example, recent concerns about global warming have politicized many so-called greenhouse gas emitting activities that previously were looked upon as environmentally unproblematic and entirely legitimate and desirable socially.

Ideational scholars stress that policy entrepreneurship is necessary because interests in protecting the environment often are ill-organized and diffuse, rather than well-organized and concentrated. According to James Q. Wilson's definition of entrepreneurial politics, which he developed in his analysis of regulation, forceful mobilization of public opinion is necessary when the costs of a regulatory policy are concentrated narrowly on, for example, one industry, while the benefits are widely distributed (Wilson 1980, pp. 357–394). Food and drug regulation and environmental regulation are typical of such policies. Because environmental protection often is a collective good that is not for the consumption of special-interest groups, those who would suffer economically because of environmental policy are more likely to organize against it than those whom it will benefit. Policy entrepreneurs therefore are critical in sensitizing potential stakeholders and mobilizing public opinion.

It is insufficient that decision makers are persuaded about the value of a policy; any environmental policy must be acceptable in the political arena. To be justifiable, an environmental policy must convincingly demonstrate that it will satisfy broader interests. Policy entrepreneurs therefore need to make convincing arguments that appeal to the beliefs, values, and interests of broader constituencies (Wilson 1980, p. 372). They may often choose to highlight the distributive, ethical, and fairness aspects of targeted environmental problems. Thus, there is much evidence from the area of environmental and social regulation confirming how "a skilled entrepreneur can mobilize latent public sentiment (by revealing a scandal or capitalizing on a crisis), put the opponents of the plan publicly on the defensive (by accusing them of deforming babies or killing motorists), and associate the legislation with widely shared values (clear air, pure water, heath, and safety)" (ibid., p. 370). Transnational coalitions of entrepreneurs similarly seek through campaigning to focus attention on societal problems, mobilize international public and political support, and pressure governments to adopt more environment-friendly policies. They politically construct targeted problems as international public "bads."

Several studies stress the importance of the concern for society's best interest when new domestic environmental policies are initiated. Based on the findings from a collection of case studies in regulatory and environmental policy making, Wilson (ibid., p. 372) has concluded that "only by the most extraordinary theoretical contortions can one explain the Auto Safety Act, the 1964 Civil Rights Act, the [Occupational Safety and Health Act], or most environmental protection laws by reference to the economic stakes involved." In the same way, with regard to health, safety, and environmental regulatory programs from the 1960s and the early 1970s in the United States, Kelman (1990, p. 40) argues that "these programs were adopted against the wishes of well-organized producers, to benefit poorly organized consumers and environmentalists."[15] Similarly, Reich (1990, p. 4) notes that many of the United States' environmental policies from the 1960s "have not been motivated principally or even substantially by individuals seeking to satisfy selfish interests. To the contrary, they have been broadly understood as matters of public, rather than private, interest. . . . People have supported these initiatives largely because they were thought to be good for *society*." According to these analysts, regulatory and environmental initiatives are in many instances intended to benefit society at large, not just individual groups. Widely accepted ideas, values, and perceptions, distinguishable from economic interests and group preferences, seem to play an important role in initiating environmental policy by shaping society's perception of how the collective interest is best served.

Policy entrepreneurs advocate ideas and spread images and metaphors that highlight the need for environmental protection and clearly identify the targeted problem. But an idea will often be resisted for political, economic, and institutional reasons.[16] Also, a new and emerging idea, as it is not yet clearly defined and is at an early stage of development, might be conceptually or scientifically weak. At other times an idea might be resisted simply because it is new. Debate and persuasion are consequently needed in order to pave the way for major policy innovations. As Kingdon (1984, p. 134) has observed, "entrepreneurs attempt to 'soften up' both policy communities, which tend to be inertia-bound and resistant to major changes, and larger publics, getting them used to new ideas and building acceptance for their proposals. Then when a short-run opportunity to push their proposals comes, the way has been paved, the important people softened up." Policy entrepreneurs therefore persistently push ideas and proposals in order to acquire influence over policy.

One way in which policy entrepreneurs support, strengthen, and protect an idea is by spreading it; another is by attacking competing or rival ideas. Evidently there is interplay between these two strategies, since entrepreneur coalitions by effectively spreading and consolidating their ideas reduce the opportunity for rival ideas to define policy problems. Indeed, as Moore has observed (1990, p. 73), "dominant ideas have certainly relegated competing definitions of the problem to relative obscurity." Moreover, a more direct, deliberate strategy of policy entrepreneurs inside governments and government agencies is simply to reduce the dissemination of competing ideas and solutions—thus attempting to monopolize the supply of ideas and solutions—by preventing rivals from participating in the making of policy. For instance, they may exclude rivals as witnesses at hearings and public meetings, as members of advisory groups, or as authors of policy reports.

Transnational entrepreneur coalitions function as selection, framing, and dissemination mechanisms for ideas. Many sources of ideas exist. Individual policy entrepreneurs may occasionally single-handedly produce powerful ideas.[17] At other times, experts and analysts supply ideas that become influential. Networks of specialists—such as issue networks, policy communities, and policy subsystems—generate and debate ideas and policy proposals.[18] By generating the ideas and proposals from among which decision makers choose, they influence the development of domestic and international policies.

Though it might still be somewhat unclear where individual ideas come from, the crucial issue is whether an idea receives sufficient support, not who or what originally produced it. Indeed, Kingdon (1984, p. 76) asserts that "ideas come from anywhere, actually, and the critical factor that explains the prominence of an item on the agenda is not its source, but instead the climate in government or the receptivity to ideas of a given type, regardless of source." It is likewise important to distinguish between the origin and the use of ideas and issues within an international institutional setting such as the European Union (Marks et al. 1996, p. 357). Influential ideas are not necessarily new ideas, and even potentially very powerful ideas often have to be framed and marketed to potential parties and stakeholders. Hence, to explain policy development it is necessary to examine both the input and the uptake of ideas—put differently, the supply of and the demand for ideas.

It is crucial to translate ideas into concrete policy proposals and coherent policy projects in order to influence policy. Common policy projects need to be fine-tuned in order to acquire the necessary political and public support and at the same time neutralize political opposition. Politicians, administrative leaders, and international bureaucrats will as individuals in a transnational entrepreneur coalition ensure that their common policy project is sufficiently compatible with political, institutional, and economic opportunities and constraints prevailing nationally and internationally. In order to transform a somewhat diffuse idea into something sufficiently concrete and viable, it is necessary that political feasibility be taken into account properly.[19] Negotiating skills as well as political and administrative experience are therefore important political resources when policy entrepreneurs target problems and attune their policy project to a given context.

For ideational scholars, the power or force of ideas depends on the intellectual as well as the contextual characteristics of ideas (Moore 1990, pp. 78–80).[20] The intellectual characteristics refer to the intellectual content and structure of an idea; the contextual characteristics refer to the historical and institutional conditions under which an idea has emerged and could guide collective action. In their intellectual characteristics, most influential public ideas are not overly complex or differentiated.[21] As Moore has observed (ibid., p. 79), "there is no clear separation of ends from means, of diagnosis from interventions, or assumptions from demonstrated facts, or of blame from causal effect. All are run together in a simple gestalt that indicates the nature of the problem, whose fault it is, and how it will be solved." Similarly, Wilson (1995, p. 256) has stressed that the power of the ideas that political elites use to influence policy often "may not be profound or well-thought-out ideas ... their power ... depends on their being plausible and satisfying representations of a new way of looking at the world." The distinctness of a public idea hinges on its ability to summarize what a particular issue is all about—or, more precisely, what it is widely perceived to be about.

It seems quite plausible, then, that it is not clear reasoning, but simple images, illustrative pictures, analogies, and even anecdotes that make public ideas powerful and influential. They make it possible for transnational entrepreneurs to draw attention to complex environmental issues and, equally important, to make sense of them in the public arena. It should therefore be expected that policy entrepreneurs especially search for and select those ideas that can effectively be expressed in metaphoric or symbolic forms. It follows that the ideas and problems that appeal most to policy entrepreneurs—in effect, the ideas and problems that policy entrepreneurs tend to select and push—are those that can be broken down into and communicated as simple and intuitively appealing metaphors, images, and symbols.[22] "The ozone hole," "Save the whales," "biodiversity," and "spaceship Earth" are some of the prominent examples of such environmental ideas and metaphors.[23] Nonenvironmental examples of policy entrepreneurs' converting complex concepts into powerful populist themes and ideas also exist.[24] For these scholars, metaphors, images, and pictures that can grab the public imagination are of utmost importance in public communication and societal dissemination of ideas.[25]

In regard to contextual characteristics, an idea, if it is to be influential, must resonate with historical experiences and must have a clear link to a currently pressing problem. As I have noted, policy entrepreneurs are most influential when they succeed in coupling their ideas and proposals to urgent problems; rarely do they have any control over the stream of problems, and seldom are they able to provoke a crisis that makes decision makers search urgently for policy solutions (Polsby 1984, p. 169). Scholars have similarly found that the persuasiveness of economic ideas is determined as much by current economic and political circumstances as by the structure of the ideas themselves (Hall 1989, p. 370). And historical studies of regulation also confirm the significance of the historical context for the power of ideas (McCraw 1984, p. 304). In summary, ideas must be fitted to existing intellectual and contextual circumstances in order to gain influence over policy.[26]

Apart from the intellectual and contextual circumstances, under what conditions should ideas and transnational entrepreneur coalitions be expected to have significant influence over regimes? This is an important question for the study of these transnational coalitions as well as any other ideas-based actor. Importantly, some ideas may raise serious distributive issues at the international level, although they do not do so domestically. More generally, it matters greatly whether regimes and policy negotiations are concerned with plus-sum or zero-sum situations. Causal ideas and persuasion are most influential when policies are concerned with efficiency (with "expanding the pie") but are unlikely to be influential in resolving distributive issues ("dividing the pie"). Indeed, Majone (1996b, p. 618) argues that "ideas matter most when collective decisions are about efficiency issues—how to increase aggregate welfare—rather than about redistributing resources from one group of society to another. Conversely, arguments are powerless when politics is conceived of as a zero-sum game."[27] Other scholars expect similarly that persuasion and joint problem solving will dominate when countries attempt to solve benign cooperation problems—that is, those problems that have potential for realization of Pareto improvements and therefore reduce incentives to cheat (Underdal 1987).[28]

Both causal and principled ideas might figure significantly in regimes concerned with the regional and the global environment, because environmental damage in principle is inflicted on all countries, creating common vulnerability as well as incentives to reduce environmental damage. In contrast, a skewed distribution of economic and environmental incentives for cooperation characterizes one-directional environmental risks and issues. A country incurs economic costs if it curbs emissions causing transboundary pollution, whereas downstream and downwind countries benefit to the extent that environmental damage is reduced. In those situations, because one part's gain is the other part's loss, causal ideas are unlikely to play a significant role. However, principled ideas attracting strong public and political support might have an impact on upstream and upwind countries.

Power-Based Regime Analysis and the Global Ocean Dumping Regime

Realist analysts and power theorists primarily study military power; they study international political economy only to a lesser extent. They have so far paid little attention to global cooperation on environmental protection. Nonetheless, it is quite obvious that realists see the structure of the international system as a severe obstacle to such cooperation. For instance, Waltz (1986, p. 106) concluded, with regard to cooperation to protect the environment: "The very problem . . . is that rational behavior, given structural constraints, does not lead to the wanted results. With each country constrained to take care of itself, no one can take care of the system."[29] A strong emphasis on how the international system forces states to protect their national interests and to ignore common interests pervades realism and power-based theory.

In the realist view, states are trapped in a static situation out of which only a major change of the international system can bring them. According to Waltz (ibid., p. 108), "states facing global problems are like individual consumers trapped by the 'tyranny of small decisions'":

> States, like consumers, can get out of the trap only by changing the structure of their field of activity. The message bears repeating: The only remedy for a strong structural effect is a structural change.

The phrase "tyranny of small decisions" refers to collectively unwanted consequences of individuals' behavior. A prominent example would be the "tragedy of the commons."[30] Numerous examples exist of how society and other "groups" may dislike the aggregate results of behavior that on the level of the individual seems rational (Schelling 1978).

Realists assert that the anarchic structure of the international system—a result of absence of a government, a police force, and a judicial power—makes nations constantly worry about their survival. In their view, the international system is a self-help system that severely constrains governments' ability to cooperate (Waltz 1986, pp. 101–104). Two principal obstacles to all cooperation follow from this: "A state worries about a division of possible gains that may favor others more than itself. . . . A state also worries lest it become dependent on others through cooperative endeavors and exchanges of goods and services." (ibid., pp. 102–103)

Following the realist approach, states would cooperate within the global ocean dumping regime only to the extent that doing so would improve, or at least maintain, their position relative to other states. States would not cooperate if the regime weakened their position relative to other states. Realists would suspect that developing countries generally would be reluctant about participating in the regime. Because of their poor economic conditions, power theorists would expect developing countries not to spend their scarce economic resources on environmental protection and pollution control technologies.[31] For similar reasons, it seems very unlikely from a realist viewpoint that developing countries would implement stringent international environmental legislation, because that would retard their much needed industrial and economic development.

For power theorists, the primary concern of states is neither maximizing power nor realizing economic gains. Instead, a state's primary concern is to maintain its position in the international system (ibid., p. 127).[32] Also for this reason, realists would doubt that developing countries would join the global ocean dumping regime. Moreover, considerations of security subordinate economic gains and all other gains to political interest. Owing to the security and energy concerns intimately associated with regulation of radioactive waste, realists would therefore predict that nuclear nations would strongly oppose other states' interference in these matters.[33] Nuclear nations would most likely attempt to exclude regulation of nuclear waste from the range of issues covered by the regime. Similarly, as long as ocean dumping of wastes is not perceived to be threatening their national security, developing countries would be very reluctant about joining and actively participating in this environmental regime.

Realists generally assert that global cooperation occurs only at the wish of major states (ibid., p. 107). They further postulate that cooperation and international order in general can only be established by a leader possessing a "preponderance of material resources"—in contemporary parlance, a *hegemon* (Keohane 1984, p. 32). To illustrate their general claim, the supporters of the hegemonic leadership theory credit British leadership for the relatively stable world economy from 1850 to 1914, and the United States for providing the leadership after World War II until the early 1970s. International crisis and disorder, then, are caused by a lack of hegemonic leadership.[34]

Power theorists will suspect that the United States constructed the global ocean dumping regime. According to both the benevolent and the coercive version of hegemonic stability theory, the United States would act on pure self-interest but ignore collective interests.[35] The hegemon, the United States, making use of its material and military supremacy, would follow a mixed strategy of material rewards, threats, and perhaps exclusion of recalcitrant governments. Through provision of side payments, or by other forms of reward, the United States would build the necessary support. Some suggest that the United States might have spent part of its revenue on building support in case it gained more than others did (Young 1991, p. 289). However, because of the relative-gains problem, some realists strongly doubt that this kind of bargaining would occur.[36]

When would a regime change occur? Power theorists generally assert that changes in the underlying relative power capabilities result in regime change. For instance, the regimes that have been established for the allocation of the radio spectrum and international telecommunications "have reflected the relative power of states and have changed as the distribution of power has changed" (Krasner 1991, p. 363). For realists, a decline of American leadership would weaken the global ocean dumping regime, and a fragmentation of hegemonic power would eventually mean regime collapse. In the absence of a hegemon, disputes would become more likely and rule violations more frequent within this environmental regime.

In view of their claim that states are "the only agents capable of acting to solve global problems," realists would expect nonstate actors—in this case, international organizations, ocean scientists, ENGOs, and prominent individuals—to be largely insignificant in the process of building and changing the global ocean dumping regime (Waltz 1986, p. 108). Moreover, realists implicitly state that the influence of nonstate actors diminishes as the number of involved states rises.

Interest-Based Regime Analysis and the Global Ocean Dumping Regime

Scholars stressing the importance of interests fundamentally share power theorists' view of the international system and regard regimes as "largely based on self-interest" (Keohane 1984, p. 57). However, neoliberals find that international rules and institutions facilitating cooperation are more widespread than what realists suggest. Interest-based hypotheses emphasize the significance of collective goods, collective action problems, mixed motives, joint and mutual gains, and plus-sum conflicts in international politics.[37] For this group of scholars, the international political system is divided up into issue areas, and states are not quite the rational, unitary decision makers power theorists would have them to be.[38] To explain interstate cooperation, neoliberals focus on regimes. They agree with the view that "actors are rarely constrained by international principles, norms, rules, or decision-making procedures"—in other words, regimes really do not matter much on their own—but they believe nonetheless that regimes, by performing a number of functions, facilitate cooperation (Krasner 1985, p. 60).

Neoliberals claim that regimes, under some conditions, help governments overcome collective action problems, especially the problems of supplying and maintaining collective goods—for example, pollution-free oceans. The Prisoners' Dilemma illustrates that governments behaving as rational egoists might fail to cooperate. Neoliberals find that regimes can improve on this socially undesirable situation. To show how, they incorporate insights from rational choice theory and microeconomics into interest-based regime analysis. They claim that as improving communication between the prisoners increases cooperation in the Prisoners' Dilemma (where the prisoners are held incommunicado), so improving communication among governments will aid cooperation. Communication is therefore crucial in coordinating the preferences of governments.

It has been demonstrated experimentally that cooperation is more likely when the Prisoners' Dilemma is played several times instead of just once because short-term gains obtained by noncooperation are outweighed by the potential benefits of cooperation in the long run (Keohane 1984, pp. 75–76).[39] In a similar fashion, neoliberals maintain that examples from international political economy support their claim that cooperation becomes more likely when governments expect that their relationships will continue indefinitely ("the shadow of the future") and retaliation is possible (Axelrod and Keohane 1986, pp. 232–234).[40] Regularizing inter-governmental relations is therefore a major function of regimes.

Neoliberals moreover argue that regimes can facilitate monitoring and enforcement by providing information on the behavior of states. Regimes will thus help in establishing the credibility of governments, especially their reputation for cooperation. Axelrod and Keohane (ibid., p. 237) summarize: "Regimes provide information about actors' compliance; they facilitate the development and maintenance of reputations; they can be incorporated into actors' rules of thumb for responding to others' actions; and they may even apportion responsibility for decentralized enforcement of rules." By stressing reciprocity among states, regimes delegitimize defection, which makes this strategy more costly (ibid., p. 250). For neoliberals, regimes thus serve a major function as gatherers and distributors of information.

Neoliberals stress that transaction and information costs rise as the number of states increases. They agree that in some cases very high transaction costs and information costs might preclude cooperation. Another threat to cooperation within groups involving large numbers of states stems from difficulties in anticipating the behavior of other states. In addition, the feasibility of sanctioning defectors is diminished in large groups, a fact that encourages free riding (Oye 1986, pp. 19–20). To improve on such situations, neoliberals point to the beneficial effect of transforming groups involving large numbers of states into collections of more manageable two-person games.[41] In their understanding, this is achieved by powerful states forcing less powerful ones to accept terms favored by the powerful ones.[42] But, since many dimensions of the ocean dumping problem are international rather than regional, neoliberals would expect countries to prefer a global regime over less attractive regional regimes.

For neoliberals, constructing linkages between issues within regimes might also increase cooperation. Regimes with many issues on their agenda—in other words, high issue density—are therefore expected to be more successful in promoting cooperation than regimes with lower issue density. Similarly, regimes are more likely to form in issue areas with high issue density than in those with lower issue density.[43] In conclusion, regimes help cooperation by providing reliable information to members, by monitoring governments, by raising the costs of noncompliance (as issues might be linked within regimes), and by pushing states to reexamine their short-term interests in the light of long-term interests.

There seems to be widespread agreement among neoliberals that regimes are built either by largely selfish hegemons supplying international public goods or by a relatively small number of sizable egoistic states.[44] Thus, similar to realists, most neoliberals would suspect that it would be necessary, because of participation of a high number of states in regime building, that the United States throw its weight behind the creation of the global ocean dumping regime. Yet neoliberals are primarily concerned with the conditions under which regimes are formed, and they favor functional explanations, so the question of *how* regimes are constructed is not really addressed.[45]

Oran Young's neoliberal model of regime formation fundamentally questions whether a hegemon is a sufficient or even a necessary condition for regime formation. According to Young, regimes are created "when self-interested parties engaged in interactive decision-making approach a problem in contractarian terms and seek to coordinate their behavior to reap joint gains" (Osherenko and Young 1993, p. 11). Young and Osherenko suggest ten distinct and at times conflicting hypotheses about how in particular environmental protection and natural resources regimes are successfully negotiated.

The first hypothesis states that participants must create mutually advantageous solutions rather than maximize their own benefits. Thus, integrative or productive bargaining behavior must prevail over distributive or positional bargaining behavior. The second hypothesis states that justice and fairness must be achieved, otherwise no regime will arise. Third, salient solutions or focal points make it more likely that regimes are created.[46] Striving for simplicity will facilitate negotiations involving numerous parties.

The fourth hypothesis states that exogenous shocks or crises may increase the probability that agreement is reached. Such events (the discovery of ozone depletion over Antarctica is one example) increase the urgency of reaching an agreement. Fifth, it is hypothesized that policy priority is essential. Two alternative formulations are suggested: that regime formation is successful only if an issue is given high priority by the participants, and that the probability of successful regime formation increases when an issue is given low priority.

The sixth hypothesis emphasizes that, in order for regime building to succeed, states must adopt a broader view of their interests and must value common goods. The seventh hypothesis reflects that environmental regimes regularly are concerned with scientific and technical issues. Three probabilistic and quite science-optimistic arguments regarding the influence of science and technical expertise on regime formation are suggested. The eighth hypothesis states that regimes will be created only if all relevant parties and stakeholders participate in the negotiation process. The ninth hypothesis suggests that the establishment of credible compliance mechanisms for regimes is an important although not a necessary condition.

The final hypothesis states that regimes can arise only if individual leaders have effective influence. Three different types of individual leaders are identified: *structural, entrepreneurial,* and *intellectual* leaders.[47] A structural leader acts on the behalf of a state and leads the bargaining process by constructive use of the power that stems from the state's material resources. An entrepreneurial leader uses negotiating skills to construct mutually beneficial solutions and to shape institutional arrangements in such as way that states are willing to accept them. An intellectual leader shapes how participants in institutional bargaining perceive the issues under consideration and think about solutions to these issues. It is suggested that entrepreneurial activities are necessary for successful regime formation, and that the interplay of at least two of these forms of leadership is required.[48] Young and Osherenko (1993b, p. 234) conclude that intellectual leaders are more important in the early stages, "usually before explicit or public negotiations begin." The consumers of ideas in this conceptual model are policy elites and negotiators. The model is not concerned with the opinion of the general public or with stable ideas enjoying broad public support. In comparison with scholars who give weight to ideas and debate in domestic and international settings, Young employs a much narrower concept of ideas and persuasion.

Neoliberals, differing with power-based theory about coercive hegemons and regime maintenance, expect a benevolent hegemon to support a regime as long as the benefits to the hegemon exceed the costs, and despite the fact that other states to some extent follow their own interests and contribute less than their share to the provision of the collective good ("exploitation" of the large by the small).[49] For neoliberals, it is therefore not necessarily the case that American interests will exclusively dictate the global ocean dumping regime's scope (i.e., the range of issues it covers) and strength (i.e., the degree of compliance with its injunctions). Other states might pursue their own interests, at least to some extent. But, similar to realists, neoliberals would expect dumpers to fiercely resist other states' attempts to reduce their autonomy regarding radwaste disposal. A successful resolution of conflict between producers (exporters) and receivers (importers) of transboundary pollution is unlikely as long as egoistic polluters have no incentive to halt their polluting activities. Increasing the malignity of the problem, the potential for cooperation on radwaste disposal is further reduced because the polluters—first and foremost the United States, Britain, France, and Japan—possess abundant economic and political power.

Knowledge-Based Regime Analysis and the Global Ocean Dumping Regime

The reflective literature stresses that cooperation cannot be fully understood without reference to ideology, to the knowledge available to actors, and to the values of actors. Cooperation is affected by ideas, by the provision of information and the capacity to process it, by perception and misperception, and by learning.[50] Ernst Haas emphasizes the role science plays in policy making within international organizations and in international cooperation, but reflectivists are also concerned with the way in which national bureaucracies make decisions and develop policies. Whereas realists and neoliberals analyze at the level of the international system, reflective scholars analyze at the level of the units of the international system.

Haas (1980, p. 357) links regime creation to perception of costs of no cooperation: "The need for collaboration arises from the recognition that the costs of national self-reliance are usually excessive." Knowledge must be rather consensual, however, in order to guide regime creation (ibid., pp. 364–367).[51] Since knowledge varies over time, reflectivists doubt that certain forms of cooperation by definition provide collective goods (ibid., p. 360).

Reflectivists do generally not consider hegemonic power a sufficient condition for regime creation and maintenance (E. Haas 1980, p. 359).[52] It is unlikely, they believe, that the hegemon will supply the necessary power in situations where it does not see the need for creating collaborative arrangements (ibid., p. 365). This might happen when those involved at the domestic level do not support cooperation, are unable to reach agreement among themselves, or miscalculate the outcomes and benefits of cooperation.[53] On the other hand, cooperation is helped to the extent governments are guided by embedded norms stressing the existence of a commonly shared problem and the appropriate solutions.[54] International organizations may play a major role as producers and distributors of new scientific knowledge. But regimes arise and change through interactions among governments, leaving little influence to other actors (for example, international organizations, individual leaders, and ENGOs). Downplaying the role of hegemons, Haas (ibid., p. 370) generalizes that "regimes are constructed by states through the medium of multilateral negotiation."

When, then, do regimes change? Reflectivists agree that changes of regimes are mainly results of change of knowledge affecting how the national interest is defined: "If we adopt this perceptual notion of the national interest, we must discard the idea of "structurally necessary" regimes; nothing is *absolutely* necessary. . . . What was considered necessary in one epistemological perspective becomes obsolete in another." (ibid., p. 392)

To understand the evolution and change in governments' perception of cooperation, reflective scholars focus much on how scientists and policy experts influence policy. They stress that professional groups (in particular, those within which scientists and experts interact) may, by way of academic training and professional experiences, acquire a common outlook on the world, and may even share political values. Similarities exist between epistemic communities and issue networks, policy communities, and policy subsystems, but reflectivists argue that an epistemic community chooses to advocate policies only when the policies conform to their causal and principled beliefs. Moreover, epistemic communities are actively formulating and influencing policy; they are not just government advisors, policy specialists, or experts.

According to knowledge-based analysis, knowledge is transformed into policy through a piecemeal, gradual process (ibid., p. 369). But it is difficult to pin down exactly when knowledge is consensual enough for policy-making purposes, and it therefore becomes difficult for the reflective scholar to predict at what point new knowledge will give rise to regimes.[55] Nonetheless, negotiations usually deal with topics on which there is an accepted pool of knowledge (ibid., p. 370). It should therefore be expected that the global ocean dumping regime and global radwaste disposal policy reflect consensual or nearly consensual knowledge.

To illustrate his thesis, Haas has often referred to ocean affairs. Pointing to marine protection as an example of a significant conceptual change, he writes that "the ocean matters to governments because their citizens use it to fish, sail ships, extract oil, fight wars, and conduct research; they also now recognize that the oceans help determine the weather and that it may not be a good idea to use them as the world's garbage dump" (ibid., p. 365). Haas does not explain, however, how and why the health of the oceans and competing uses of the oceans became significant political issues. As we shall see, both issues signify a change in perception that gained political momentum only in the early 1970s. Thus, the reflective approach often has an evolutionary and rather harmonious flavor. As my study shows, it tends to downplay ideological conflicts and rivalries among competing perceptions.

When studying environmental regimes, scholars stressing the importance of cognitive factors focus mostly on scientific and technical knowledge and on how expert advice is organized and institutionalized. Peter Haas (1990b, p. 350) argues that an ecological epistemic community—which shared views on "the kinds of substances to be controlled, the methods to be used and the values to be employed in order to direct policy towards desired ends"—has been influential in promoting the arrangement for protecting the Mediterranean Sea against pollution. This ecological epistemic community defined the scope and influenced the strength of the Med Plan. Based on a shared broad understanding of the environmental problems and their solutions, an ecological epistemic community pressured decision makers to construct a regional arrangement the community itself had defined. The strength of this regional pollution control arrangement varied in relation to the ecological epistemic community's influence on domestic policy making. The countries where the epistemic community was strongly represented, i.e., privileged access to national decision makers, were the most active supporters of international commitments and the most successful in national compliance along the lines of the epistemic community's shared view. Countries with weak representation of the ecological epistemic community were less supportive of international commitments and adopted weaker domestic pollution controls. Governments would be persuaded to establish environmental protection policies when individuals from the ecological epistemic community presented their advice in a forceful and consistent way to national decision makers. This also served to minimize the influence of the opposition.[56] Similar to the case of the Med Plan, Haas (1990b, p. 361) claims that recent international efforts to protect the ozone layer were driven largely by an ecological epistemic community that "identified the broad scope of international policy" and "pressured governments to comply with international standards." Litfin (1994), however, has concluded that epistemic communities and consensual knowledge as defined by Haas in reality were not so influential in the development and framing of the international response to the stratospheric ozone depletion problem.

Epistemic-community theorists argue that the political influence of an epistemic community grows with its control over bureaucratic power. In countries in which it gains significant bureaucratic control, it will institute environmental protection policies. At the national level, members of an epistemic community might be present in areas such as budgetary finance, staffing, and enforcement authorities (Haas 1990b, p. 351). At the international level, an epistemic community might supply international officials who will influence agenda setting and policy debates and who will draw attention to international problems and their possible solutions. Thus, an epistemic community might, as a transnational network, influence policy making in several countries.

The epistemic-community analyst would be inclined to look for evidence that an ecological epistemic community basically created the global ocean dumping regime. The ecological epistemic community would persuade domestic decision makers of the need for protection of the marine environment against ocean dumping. The epistemic community would press for a regime that reflected its perception of the scope of this problem. Those countries in which the epistemic community had access to domestic decision makers would be the strongest supporters of stringent controls on ocean dumping. Conversely, those countries in which it had limited or no access to domestic decision makers would be the weakest supporters of stringent controls on ocean dumping.

Although Haas focuses on the role of knowledge and new information when explaining why Mediterranean states complied with and expanded the Med Plan, he actually points to erroneous beliefs instead of scientific knowledge when he explains the establishment of the regime: "Many [national] officials thought that pollution was a commons problem, and thus required coordinated action throughout the region. They assumed that currents transferred the pollutants fairly freely among countries. UNEP officials were well aware that currents were not sufficiently strong to transmit pollutants across the Mediterranean Basin . . . but they hoped to complete an agreement, so they just smiled and nodded when others characterized Mediterranean pollution as a commons problem. . . . This false perception actually facilitated the resolution of the problem." (Haas 1990a, pp. 70–71) Paradoxically, therefore, persuasion by an ecological epistemic community apparently was not necessary in bringing decision makers to begin controlling regional pollution.

Elsewhere, Haas briefly mentions that "Jacques Cousteau was active in attracting publicity" to Mediterranean pollution, and that "gloom-and-doom prophesies" produced widespread concern about the state of health of the Mediterranean in the early 1970s (1990a, pp. 83, 104). Moreover, Haas occasionally juxtaposes the influence of popular science and hard science in his explanations. He writes, for example: "An ecological epistemic community was consulted by governments in order to dispel uncertainty about the extent of environmental pollution. Such concern was precipitated by a crisis; the alarm that the Mediterranean was in danger of dying." (ibid., p. 224) Yet, the fact that prominent scientists and ecologists together with popular science accounts of Mediterranean pollution identified the need for regional environmental cooperation is not taken into account by the epistemic-community approach.[57]

To create the global ocean dumping regime, the epistemic-community analyst would expect that an ecological epistemic community would supply international leadership. Domestic decision makers would be uncertain about the nature and scope of the ocean dumping problem, as well as the costs of specific control strategies, and would consult the ecological epistemic community for advice. By providing the authoritative and legitimate scientific understanding of the ocean dumping problem, the ecological epistemic community would shape policy as well as institutional and organizational arrangements. The epistemic community would develop a coherent view and common understanding of this environmental problem through international scientific conferences as well as other forms of exchange and communication of scientific understanding and knowledge. Common understanding among the members of the ecological epistemic community would guarantee that its members in various countries would give consistent, uniform scientific advice to decision makers. Reflecting the beliefs of the ecological epistemic community, domestic decision makers would then design and implement a global policy to protect the oceans against dumping.

These scholars maintain that an epistemic community's political influence remains unabated as long as it can advise policy makers in a consistent and persuasive way, and as long as its claim to expertise remains unchallenged. Its political influence might end, however, should new *Torrey Canyon*–size oil spills, ozone depletion over Antarctica, Chernobyl nuclear accidents, or shifts in scientific understanding result in rejection of the paradigm advanced by the epistemic community (Haas 1990b, p. 353).[58] In the absence of ecological or environmental crises, or erosion of its authoritative claim to policy-relevant knowledge, it thus is expectable that the global ocean dumping regime, through advice and pressure, would largely develop consistently with the interests of an epistemic community. But reflective scholars and epistemic-community theorists downplay the significance of public opinion and environmental groups in their understanding of cooperation on environmental protection.

4
Scientific Advice and Ocean Dumping: Knowledge-Based Regime Analysis

Since the ocean dumping problem gave rise to a global regime, reflectivists would be inclined to believe that in 1972 most states perceived ocean dumping as a global commons problem and that the negotiations on the regime would proceed as a largely cooperative process. This was not the case, however, as Russell Train, head of the U.S. delegation to the London Convention negotiations, explained before the United States Senate Committee on Foreign Relations in the spring of 1972: "Perhaps naively, I had thought everybody would be in favor of doing something effective about stopping dumping in the ocean, but we have found that many of the [less developed countries] are very leery of getting into this. They would rather not have a convention." (Train 1972, p. 17)

The proposal to establish a global ocean dumping regime was made by the United States. Reflective scholars would therefore examine whether scientists and experts had persuaded U.S. decision makers to propose that states work together in controlling ocean dumping. This group of theorists would assume that decision makers, who would be uncertain about the nature of the ocean dumping problem, its possible solutions, and costs of possible control strategies, would consult and possibly take advice from marine biologists, ecologists, and other scientists serving as policy experts. Reflectivists would expect that decision makers would then transform consensual or nearly consensual knowledge and expert advice into stringent control of ocean dumping.

Scholars stressing the importance of technical and scientific knowledge would, in addition, investigate whether marine scientists and other experts would constitute an epistemic community providing decision makers with nonconflicting scientific advice. Epistemic-community theorists would examine whether international expert advice and pressure had influenced

U.S. ocean dumping policy in 1972. One possible source of influence would be international scientific conferences; another would be the many specialized agencies of the United Nations that had been active in the marine pollution field since the mid 1960s.

The United States' Response to Ocean Dumping

In the late 1960s, asbestos, DDT, smog, and a few other names of supposedly deadly chemicals and environments became household words in the United States.[1] Together with nuclear fear, which had a longer history of raised emotions, these were the subject of the most frequent protests in the "pollution battle."[2] They each could serve as a guide to the surge of environmental policies that in 1969 started to sweep the United States. This period witnessed the passage of the National Environmental Policy Act (1969), the Occupational Safety and Health Act (1970), the Clean Air Act (1970), and the Federal Water Pollution Control Act Amendments (1972). Furthermore, President Richard Nixon established the Council on Environmental Quality in 1970, and the Environmental Protection Agency was created the same year.

The late 1960s was a time of dramatic and rapid changes of beliefs and values in the United States. In its issue of January 26, 1970, *Newsweek* reported, in an article titled "The Politicians Know an Issue," that "Old Washington hands have been sensing for some time that environment may well be the key issue of the '70s, for the nation and for their political futures. They freely concede that no other cause has moved so swiftly from the grass roots into the arena of public policy making. As early as 1968, environment had surpassed law and order and in 1969 was gaining on Vietnam in total linage in the *Congressional Record*. And by now, nearly everyone on Capitol Hill seems to be actively against pollution, causing a veritable stampede for stage center in the crusade to save America's land, air and water." In 1973, the environment had clearly established itself as an issue. "A few years ago," Rosenbaum noted (1973, p. 51), "'ecology' and 'environmental protection' did not exist in American public discourse; they were nonissues to most citizens and public officials. Now we are in the midst of a new 'environmental decade.'"

Faced with unprecedented demands for political action to protect the environment, President Nixon was under pressure to act. "There is no doubt," according to *Newsweek* ("The Politicians Know an Issue"), "that the President has been a Johnny-come-lately on the environmental bandwagon." Whereas Senator Edmund Muskie (a Maine Democrat who was seen as a likely presidential candidate) had made opposing pollution a central part of his political program, Nixon's position on environmental control was unclear in early 1970. "What the issue needs," said one liberal Democrat, "is a man like Bob Kennedy who went to the Indian reservations. Some young guy has to get into a bathing suit and jump into Lake Erie. If he survives, he's on top of the issue." ("And How Are the Democrats Doing?" *Newsweek*, January 12, 1970)

Pollution of Lake Erie had emerged as one of the major environmental issues in the United States, often seen as symbolizing the political crisis of the environment. According to the most widely quoted environmentalists in the United States, Lake Erie had "died" in the mid 1960s as a result of a greatly accelerated eutrophication (lake aging) process caused by industrial pollution.[3] Many saw the condition of the lake as at least gloomy, at worst doomed.[4] According to *Newsweek*'s "Special Report: The Ravaged Environment" (January 26, 1970), "a few years ago—nobody was paying close enough attention to tell exactly when—Lake Erie died: acidic wastes from the surrounding factories have strained its water of virtually every form of life except a mutant of the carp that has adjusted to living off poison." As a conspicuous symbol of industrial destruction of ecosystems of large bodies of water, Lake Erie served as a catalyst for the realization of U.S. dumping policy for both the Great Lakes as well as coastal waters (interview, Charles Lettow, September 24, 1991).

President Nixon gave a series of speeches in which he laid out a philosophy and a policy for protection of the environment. In his State of the Union message in late January of 1970, Nixon declared: "The great question of the seventies is, shall we surrender to our surroundings, or shall we make our peace with nature and begin to make reparations for the damage we have done to our air, to our land and to our water?"[5] Nixon's special "Message on the Environment" of February 10, 1970, the most comprehensive statement ever made by an American president on the subject, established a permanent three-member White House Council on Environmental Quality.[6] Although Nixon had opposed the proposal at first, the Environmental Protection Agency was also established.[7] In a message to the Congress on the subject of Great Lakes and other dumping

(April 15, 1970), the president directed the Council on Environmental Quality to study and report on ocean disposal of solid waste.[8]

In October of 1970, the Council on Environmental Quality announced a national policy to prevent pollution of the oceans in the future. The CEQ also recommended that the United States take the initiative to establish international cooperation on ocean dumping, and that such proposals should be presented in "international forums such as the 1972 UN Conference on the Human Environment" (CEQ 1970, p. 37). Such proposals should be made along the lines of national policy. President Nixon welcomed the council's recommendations and its approach of "acting rather than reacting to prevent pollution" (Smith 1970). Senator and "antipolluter" Gaylord Nelson, the Senate's leading authority on ocean pollution, declared: "We have the opportunity now to prevent the sea from becoming the same kind of mess we now see in our rivers."[9] A *New York Times* editorial titled "To Save the Seas" (October 13, 1970) cautioned that "if the nations continue to use the seas as a dump, scientists warned last summer, the world's oceans will become as "dead" as Lake Erie by the end of this century."

In a report titled "Ocean Dumping—A National Policy," the Council on Environmental Quality urged that strict limits be set on the wastes that until then had been dumped indiscriminately into the Great Lakes and the oceans. On radioactive wastes, the report said: "The current policy of prohibiting ocean dumping of high-level radioactive wastes should be continued. Low-level liquid discharges to the ocean from vessels and land-based nuclear facilities are, and should continue to be, controlled by Federal regulations and international standards. The adequacy of such standards should be continually reviewed. Ocean dumping of other radioactive wastes should be prohibited. In a very few cases, there may be no alternative offering less harm to man or the environment. In these cases ocean disposal should be allowed only when the lack of alternatives has been demonstrated. Planning of activities which will result in production of radioactive wastes should include provisions to avoid ocean disposal." (CEQ 1970, pp. vi–vii)

The Council on Environmental Quality's decision was based on the anticipated environmental and economic consequences of future dumping. A member of the CEQ explained before a Senate hearing: "We think it is a serious problem today. It is potentially a very critical problem for the future.

And let's stop; let's control the problem. Prevention, I think, in this case will be far more economical than trying to cure it after it becomes critical."[10]

In 1968, more than 48 million tons of waste were dumped at 246 sites in the Gulf of Mexico and the Atlantic and Pacific oceans. Ocean dumping was expected to increase drastically as a result of a future increase in the U.S. coastal population. The CEQ report stressed the need for preventive rather than remedial action. Knowledge of economic costs and environmental impact of ocean dumping was scanty, however, and comparisons made with disposal alternatives disregarded political and social hindrances and hoped for technological advances and new methods of recycling.

At subsequent hearings, waste managers pointed to a possible future scarcity of suitable landfills and a lack of treatment technologies for certain toxic and other hazardous materials.[11] The CEQ report listed as "interim alternatives" hauling waste to suitable dumping sites by rail and reclaiming strip mines and other land for dumping. Permanent alternatives to ocean dumping, which neither the Council on Environmental Quality nor legislators gave much attention to later, would be sought through research authorized as part of legislation and development of environmentally acceptable and feasible land-based alternatives.[12] President Nixon practically quoted the CEQ's findings and policy recommendations in his statement to the Congress: "In most cases, feasible, economic, and more beneficial methods of disposal are available. . . . Legislation is needed to assure that our oceans do not suffer the fate of so many of our inland waters, and to provide the authority needed to protect our coastal waters, beaches, and estuaries."[13]

Although the draft bill of the national regulation was prepared and cleared through the Office of Management and Budget in October of 1970, it was held and announced in the president's environmental message to the Congress on February 8, 1971 (Miller 1973, p. 59). In the Message on the Environment, President Nixon recommended legislation to implement "a national policy banning unregulated ocean dumping of all materials and placing strict limits on ocean disposal of any materials harmful to the environment" ("International Aspects of the 1971 Environmental Program: Message from President Nixon to the Congress," p. 254). In order to complete national legislation, Nixon said, the United States should work toward getting other nations to adopt and enforce similar measures.[14] The final act similarly said that the Secretary of State and the administration should pursue international action and cooperation to protect the marine environment and "may, for this purpose, formulate, present, or support specific proposals in the United Nations and other competent international organizations for the development of appropriate international rules and regulations in support of the policy of this Act."[15]

Two days later, EPA Administrator William Ruckelshaus submitted to the Senate a draft of a bill for national legislation. The bill was part of a comprehensive and wide-ranging action program that built on the 23 legislative proposals and the 14 acts to clean up air and water submitted to the Congress the year before. "Upon introduction in the Congress it was clear," wrote one of the staff members to the hearings, "that the issues would be primarily jurisdictional, not substantive: a political demand for the legislation was evident; the regulatory techniques chosen were acceptable without objection." (Miller 1973, p. 59)

The Scientific Basis of Ocean Dumping Regulation

Ocean scientists presume that the oceans have an identifiable assimilative capacity.[16] This concept relies on the capacity of the oceans to absorb and neutralize pollutants. It follows that as long as this assimilative capacity is not exceeded the marine environment will clean itself. Consequently, pollution occurs when a certain marine capacity is exposed to pollutants that exceed the upper level or capacity of assimilation of contamination. In this definition room is given to a certain legitimate use of the waste disposal capacity of the oceans, as long as the regeneration of the ocean resources is not prevented. It is therefore necessary to consider specific conditions—such as mixing capacity, turnover time (it takes, for example, 25–40 years for the water in the Baltic Sea to be renewed), stratification of water, temperatures, and level of biological activity—when defining the assimilative capacity of a certain region. As used by professionals, the word *contamination* signifies what is less than clean but not quite polluted.[17]

In the early 1970s, knowledge of environmental effects of ocean dumping was at best rudimentary. In 1971, R. B. Clark, British scientist and editor of *Marine Pollution Bulletin*, described the state of the art in marine pollution research as follows: "Most knowledge of the biological consequences of marine pollution is derived from studies in temperate waters. Information about these environments is woefully inadequate, but it is encyclopedic compared with what we know about even the basic ecology of

Arctic and tropical waters, let alone the consequences of effluent disposal and accidental pollution in them." (Harwood 1971) Two years later, it was obvious that there still was a great gap in scientific knowledge about pollution of the seas. "Despite broad general consensus about the contaminants which represent the greatest threat of global pollution," said the head of the U.S. National Oceanic and Atmospheric Administration in 1973, "knowledge of their full extent, their fate and their impact on the ocean ecosystem is scanty at best" (Newman 1973).[18] This lack of knowledge did not, however, dampen regulatory policy making.

The question of whether available knowledge of water and marine pollution could guide, and should guide, policy making had already been raised before the CEQ issued its report on ocean dumping. The scientific basis of a proposal to control waterways, included in the 37-point program presented at the first "Message on the Environment" in February of 1970, had been received with skepticism. Under this plan, rivers and lakes were assumed to have a capacity to absorb waste without becoming polluted (technically known, as already mentioned, as the assimilative capacity), and through fair allocation of this capacity among all industrial and municipal sources precise limits on the amount of waste dumped into a river or lake would be assigned.[19] But the Nixon administration's proposal, which allowed waste discharges but would avoid pollution, did not have the full consent of the Congress. Influential legislators disagreed with, in Representative John Dingell's words, "some of the industrial and municipal folks who have ideas we should utilize the streams and lakes and oceans up to their assimilative capacity."[20] The newly created EPA did not intend to base its policy on the assimilative capacity concept and seemed generally very reluctant about market-based and other approaches that might have been interpreted as favoring selling "licenses to pollute" (Wilson 1980, p. 376).[21] Concerns were also expressed about the state of the art in this field. "We just don't know enough about a river's assimilative capacity," objected one Senate water expert. "The best route is no dumping at all, but that's not what Nixon seemed to say." ("Pollution: The Battle Plan," *Newsweek*, February 23, 1970)

The CEQ report on ocean dumping actually acknowledged that existing knowledge of ocean pollution was either rudimentary or, in fact, did not exist.[22] It had, in addition, been impossible to separate the effects of ocean dumping from the broader issue of ocean pollution. The Council on

Environmental Quality nonetheless concluded that there was "reason for significant concern" (CEQ 1970, p. 18). Moreover, it was clear to U.S. government officials with experience in control of oil pollution from tankers that international control of ocean dumping was needed (interview, Charles Lettow, September 24, 1991).

Quite incompatible with the propositions of the epistemic-community approach, congressional hearings held in spring 1971 demonstrated that experts disagreed whether significant pollution had occurred in the ocean, whether an ocean had the capacity to safely absorb some wastes, and whether regulation reducing ocean dumping was justified. The view of the Nixon administration and a group of congressmen backing ocean dumping legislation and the view of representatives from the waste management field were strongly at odds. Professional witnesses from the waste management field did not find that waste disposal necessarily was a danger to the health of the oceans. They stressed that the oceans were robust, had an enormous capacity to receive waste safely, and should be considered in any rational waste management strategy. Indeed, they did not support stringent ocean dumping control, and many strongly disagreed that ocean dumping should be advised against in all cases. But another group of witnesses, mainly ecologists and marine scientists, disagreed. They advocated stringent control of ocean dumping on the ground that irreversible damage otherwise would be inflicted on the oceans. In their view, the assimilative capacity of the oceans was limited and should be protected by legislation. The lack of any form of dumping regulation resulted in a lack of incentives to reduce the amount of waste disposed of.[23]

Contradicting an important implicit claim of the epistemic-community approach, public opinion was far from indifferent on the issue of ocean pollution. Further illustration of the wide divergence of views among the acknowledged experts and, at the same time, of considerable public concern over the health of the oceans, is found in a testimony given by David D. Smith, a marine geologist who had co-authored the study that formed the basis of the influential CEQ ocean dumping report. While he advised against any bill that ignored the assimilative capacity of the oceans, Smith realized that the examination of the pros and cons of ocean dumping was met with public disbelief. Clearly, public opinion was in favor of significantly reducing, if not completely banning, ocean dumping of wastes. Smith told the Senate: "We are faced with a matter of attitude in the United States today. It seems clear that in the general public's mind the idea prevails that disposal of any waste materials in the ocean is inherently bad, and therefore should be stopped, or at least severely curtailed. . . . I am convinced, and I believe if you will talk to various professionals in the waste management field you will find general agreement, that ocean dumping of selected types of waste—and I emphasize selected—is not only permissible but is in fact quite desirable."[24] Smith concluded his remarks by stressing the assimilative capacity of the oceans: "There is a need to recognize in the bill . . . that the waste assimilative capacity of the sea is enormous. I can hardly overstate or overemphasize that there has been a general failure to recognize this. We hear a lot of what I term vastly oversimplified and commonly ill-founded statements that any discharge of waste to the sea is pollution. This is just not true. If you will talk to qualified sanitary engineers, qualified biologists who are concerned with waste management, I think you will find general agreement with this."[25] Parts of this witness's statement were reprinted in the Senate report (which, however, ignored it and instead urged strict control on ocean dumping).[26]

Despite a lack of knowledge about the environmental effects of ocean dumping, Congress passed a law on ocean dumping regulation the following year: the Marine Protection, Research, and Sanctuaries Act of 1972. It was hoped that research would shed light on the many unknowns of ocean pollution. Motivation to solve the problem was obviously far ahead of understanding.

Also contradicting the epistemic-community approach, chapter 5 will document that a group of legislators seeking to minimize all ocean dumping mobilized public and political support for U.S. domestic legislation. To supplement domestic regulation, they also advocated the establishment of a global regime on ocean dumping. To influence and move public opinion and political leaders, both national and international, a series of congressional hearings spread the simplistic, powerful idea that "the oceans are dying." Although clearly disregarding the many scientific unknowns of ocean dumping, this idea was successfully used to focus public opinion and legislative attention on this issue.

The Pre-Stockholm International Response to Ocean Dumping

As epistemic-community theorists and other theorists predict, in the early 1970s international officials saw in protection against ocean dumping and in ocean protection generally a way to expand their sphere of competence and influence.[27] Several United Nations specialized agencies—in particular the Intergovernmental Maritime Consultative Organization (predecessor of the International Maritime Organization), the Food and Agriculture Organization (FAO), the United Nations Educational, Scientific and Cultural Organization (UNESCO), and the World Meteorological Organization—had been involved in various aspects of ocean pollution since the 1960s (Gardner 1972).[28] In the words of a U.S. Department of State official: "A number of international organizations—in fact, I dare say almost every one of them—all simultaneously discovered the environment. All decided that they in turn want to be the sole organization or the principal organization dealing with it."[29] A pronounced tendency among international public officials to make ocean pollution, as well as the environmental "crisis," a serious international issue was also obvious.[30] But none of these international agencies had success in getting governments involved in ocean dumping control. Neither did they play an influential role in the construction of the global ocean dumping regime.

UNESCO's so-called Biosphere Conference, held in Paris in September of 1968, marked the beginning of the new era of international environmental concern. The official name of this conference was "Intergovernmental Conference of Experts on the Scientific Basis for the Rational Use and Conservation of the Resources of the Biosphere." The conference was organized in cooperation with other UN organizations and with several nongovernmental agencies, including the International Union for the Conservation of Nature and Natural Resources (an organization aimed at furthering the ecological point of view).[31] Earlier conferences on environmental issues, often sponsored by the United Nations, had tended toward a technical rather than an ecological orientation and had been single events.[32] The Biosphere Conference concentrated on the scientific aspects of conservation of the biosphere and marked the first appearance on the international environmental agenda of the biosphere approach to human-environment relationships. The conference was well attended; representatives from 63 nations and a number of international organizations were present. As at the 1972 United Nations Conference on the Human Environment (the so-called Stockholm conference), prominent and vocal ecologists, mostly American, attended and contributed to conference reports.

The final report of the Biosphere Conference included this recommendation: "In the place of single-purpose actions in disregard of their associated consequences, both public and private, there is need to substitute planned programs for the management of resources if past degradation of the environment and deterioration of ecosystems are to be corrected, if the biosphere's productivity is to be maintained and even enhanced, and if aesthetic appreciation is given opportunity to flower." (UNESCO 1970, p. 234) But because the economic, social, and political dimensions of the problems of the biosphere were outside the purview of the Biosphere Conference (UNESCO is concerned primarily with science issues and scientific aspects of policy making), the final report made only vague recommendations for future legal and institutional changes. It concluded (ibid., p. 235) that "it has become clear . . . that earnest and bold departures from the past will have to be taken nationally and internationally if significant progress is to be made," but the more precise nature of those "departures" was not mentioned. The 1972 Stockholm conference, in contrast, would focus on the economic, political, and social dimensions of protection of the global environment.

"Man and His Environment: A View Toward Survival" was the telling title of the 13th National Conference of the U.S. National Commission for UNESCO dealing with the environment. The conference took place in November of 1969 at Stanford University and was attended by representatives of more than 200 organizations. Prominent environmentalists such as Paul Ehrlich, Barry Commoner, and Margaret Mead contributed background papers or participated in panel discussions, or both. Considerable attention was paid to ocean pollution as well as to control of population growth, reduction of atmospheric pollution, and preservation of ecological diversity. One scientist warned that "the end of the ocean came late in the summer of 1979 and it came even more rapidly than the biologists had expected" (Pryor 1970, p. 115).[33] Another contribution (The Sea: Should We Now Write It Off As a Future Garbage Pit?) was headlined as follows in an anthology (Riseborough 1970, pp. 121–136): "For those who don't as yet believe that the sea is dying, this is ample proof. For those who do, it is further documentation." It was concluded that "scientific, practical, economic, moral, and esthetic reasons require that the sea *not* be used as a garbage dump" (ibid., p. 122). Much as had happened at the Biosphere Conference, it was concluded that international machinery was needed because DDT, polychlorinated biphenyls,[34] and radioactivity released by atomic explosions were capable of traveling great distances. The conference proposed that "the leaders of all nations through the United Nations

General Assembly declare that a state of environmental emergency exists on the planet Earth."[35] Among the proposals for future action were the establishment of national, regional, and worldwide commissions on environmental deterioration and rehabilitation. However, more precise indications were not given.

The beginning of the United Nations' interest in international control of marine pollution was signaled by a General Assembly request from 1969 that the Secretary-General conduct a survey among member states on desires for international arrangements for regulation and reduction of ocean pollution.[36] The responses from 44 countries showed a general concern with the increasing threat of pollution to the ocean environment and with the need for international prevention and control of ocean pollution. It was reported that no existing international agreement effectively controlled marine pollution. Existing agreements were too broad, and there was no proper enforcement of many of the concepts agreed to. Other agreements were narrow and did not cover the existing range of pollution problems (Schenker 1973, p. 41).

The UN's more specialized agencies also involved themselves in ocean dumping. In 1969, the Intergovernmental Maritime Consultative Organization conducted a survey among its member countries on the kind and amount of materials which were disposed of from ships and barges in the ocean (IMCO Document OPS/Circ.15, May 13, 1969, also appendix to GESAMP report no. 22, 10 February 1970). The general picture showed that some control of dumping existed within territorial waters but that there was almost no control of dumping on the high seas (Böhme 1972, p. 105).

In December of 1970, the FAO organized "the first attempt to make a worldwide scientific approach to marine pollution and its effects on the living resources of the oceans" (ibid., p. 91). The Technical Conference on Marine Pollution and its Effects on Living Resources and Fishing was held in Rome from December 9 to December 18, 1970. Invitations had been sent to all FAO member nations and associate members, to UN agencies, to intergovernmental organizations, and to nongovernmental bodies interested in ocean pollution. Almost 400 individuals attended the conference, most of them scientists and experts in the various fields of marine pollution but some from industry or government. They came from more than 65 countries. In addition to its scientific objectives, the conference intended to focus attention on pollution problems where international cooperation and coordination were required. The experimental and review papers were mostly concerned with pollution of rivers and atmospheric fallout. Ocean dumping received little attention. In the discussion it was pointed out, however, that the future impact of marine pollution on a large scale would derive from ocean dumping. Serious concern should therefore be given to the future development of disposal of waste by ocean dumping. Harmful substances were reaching the ocean from coasts, through rivers and the atmosphere, but the substances disposed of by ocean dumping—radioactive materials, chemical weapons, and ammunition—were particularly toxic and persistent. The participants concluded therefore that "in future there is all the more an urgent need to improve the knowledge and information about the aspects of pollution by ocean dumping before any future control of ocean dumping can work efficiently" (ibid., p. 96). The recommendations said that the FAO, in cooperation with other bodies, should "review the widespread practice of dumping wastes, especially toxic or persistent substances in the world oceans and encourage international studies of selected dumping sites to make a scientific evaluation of both the short and long-term effects of such practices, and bring about cessation of the practice of dumping containers of waste and other obstacles in present and potential fishing grounds, and establish a system of registration to cover the dumping of all persistent and or highly toxic pollutants into the sea" (ibid., p. 95). But the FAO was not involved in the preparations already underway to establish a regional arrangement for protection of the North Sea against ocean dumping (the Oslo Convention), and it did get only indirectly involved in the preparations for the global ocean dumping regime, initiated only a few months after the marine pollution conference.

Conclusions

Quite contrary to propositions made by scholars stressing the influence of scientific and technical knowledge on decision making, the Nixon administration proposed regulation of ocean dumping, although scientific evidence of damage to the oceans was almost nonexistent in the early 1970s. Despite a lack of knowledge, the Nixon administration considered ocean dumping a domestic environmental problem of some importance and urgency. Moreover, because of ocean dumping's international character, the administration suggested that international regulation be established. Although some epistemic-community theorists doubt that leadership by prominent states is necessary for regimes to form, the hegemon, the United States, thus took the initiative to establish the global ocean dumping regime.

Nor were politicians influenced by scientific knowledge, as epistemic-community theorists suggest. Congressional hearings demonstrated that scientists radically disagreed as to whether ocean dumping had damaged the ocean environment and whether the ocean had the capacity to safely absorb some substances. Scientists also disagreed as to whether knowledge about the effects of ocean dumping was sufficient to guide regulation. In short, technical and policy experts were divided. Stringent ocean dumping legislation was, therefore, not given unanimous support by marine scientists. Politicians, however, largely ignored this.

Beginning in the late 1960s, international organizations, scientists, and international "anti-pollution" conferences advocated control of ocean dumping. Ecologists and environmentalists participated in conferences (for example, the Biosphere Conference) organized by international organizations; ecologists and environmentalists also participated in international "anti-pollution" conferences; and the FAO convened several hundred scientists to establish a global scientific approach to marine pollution. Specialized UN agencies began collecting information on national dumping regulations and on the amounts and the kinds of wastes being dumped. But the public and the governments paid little attention to such initiatives.

In chapter 5 I will show that politicians together with prominent environmentalists and ecologists—instead of scientists and international organizations—spearheaded the U.S. initiative to construct a global ocean dumping regime. Perhaps epistemic-community theorists would object to this conclusion. They would correctly point out that the global ocean dumping regime of 1972 was constructed at a time when few governments had established environmental protection agencies. An epistemic community would, therefore, lack the organizational platform necessary to exert its influence. Nonetheless, this case demonstrates that the existence of an epistemic community is not a necessary condition for environmental regime formation. Chapters 5 and 6 will add supporting evidence to this conclusion.

5
Ocean Dumping and U.S Domestic Politics: Power-Based Regime Analysis

Realists (and neoliberals) suspect that the United States constructed the global ocean dumping regime so as to use it to realize its environmental objectives—at the expense of other states, when that was necessary. They see the regime as closely mirroring the United States' environmental interests.

Realists view the global ocean dumping regime as perhaps satisfying other foreign policy objectives of the United States, such as improving its bargaining position in international trade or strengthening an international organization facilitating cooperation. These foreign policy objectives are not intrinsically or primarily related to the solution of environmental problems. A 1970 U.S. administration study report suggested: "International cooperation on the environment may be deliberately undertaken and encouraged for the purpose of strengthening an international organization that serves U.S. interests, revitalizing it, and/or to increase its capability to bring nations together. Sometimes the primary aim is a more specific political objective such as enhancing the United States' image abroad or improving our bargaining position in international trade. In these cases, the improvement of environmental quality is a secondary consideration even though the desirability of that end is acknowledged and the need for joint action to control pollution is recognized."[1] These theorists assume that the United States dictated the terms of the regime to other states through a combination of coercion, co-optation, and manipulation of incentives.

In order to assess the realist approach to regime formation, this chapter examines the formulation of U.S. domestic policy, influential politicians' perception of the ocean dumping problem, Congress's support of legislation, and the economic implications of U.S. domestic ocean dumping regulation. Contrary to realist propositions about the crucial weight states and hegemons give to egoistic self-interest, this chapter shows that ocean dumping was perceived as an international and even a global environmental problem by Congress and by the Nixon administration in the early 1970s. Through a series of hearings intended to attract public attention and to encourage legislative action, a group of legislators who claimed that they were "trying to clean up the oceans"[2] established domestic regulation. To influence and move public opinion and political leaders, nationally and internationally, congressional hearings spread the simple, powerful idea that "the oceans are dying." To supplement domestic regulation, it was argued, the United States should also work toward agreement on a global ocean dumping regime.

Supporting the claim made by realists and power theorists, introduction of ocean dumping regulation domestically created significant pressure in the United States for some sort of global regulation able to harmonize the economic costs of environmental protection across countries. But the realist proposition that the United States would be motivated solely by egoistic self-interest and a need to vigorously protect the national interest does not conform well to this case. Political leaders wanted the United States to persuade other countries to follow by setting a good example and demonstrating willingness to act against ocean dumping.

Congressional Hearings on Ocean Dumping

The U.S. Department of the Army's disposal of some 65 tons of nerve gas in the Atlantic Ocean off Florida in the summer of 1969 focused national and international attention on the problem of unregulated ocean dumping of supposedly extremely dangerous materials.[3] In the words of one congressman (*Congressional Record: House*, September 8, 1971, p. 30854): "The nerve gas dumping incident reverberated around the world and focused public opinion on the need for legislation."[4] In August of 1970, despite national and international protests, the Army disposed of surplus nerve gas rockets embedded in concrete vaults on the ocean floor deep under international waters.[5] "A major incident," wrote Robert Smith (1970) in the *New York Times*. Concerned about this dumping practice, the UN Seabed Committee made an after-the-fact "appeal to all governments to refrain from using the seabed and ocean floor as a dumping ground for toxic, radioactive and other noxious material which might cause serious damage to the marine environment" (Deese 1978, p. 47). Scientists concerned over plans for oil drillings and discharging of domestic wastes, chemicals, minerals and "other byproducts of our technology by proposed giant outflows into the deep sea" (Howard Sanders before Subcommittee on Oceanography, quoted from Senate Report no. 451, p. 4238) gave testimony before congressional hearings describing possible catastrophic implications in the deep sea and were quoted in the press.[6] A 1969 study by the National Academy of Sciences talked of catastrophic dangers for fish. In October of 1970, when the Council on Environmental Quality announced its ocean dumping report, the Army's dumping was again brought up. "Such practices could—and should—be controlled by executive order to conform to the new guidelines," editorialized the *New York Times* ("To Save the Seas," October 13, 1970).

Public and political attention to ocean dumping was sustained through several congressional hearings held in 1971. A group of representatives and senators concerned over ocean dumping, some with ties to the U.S. marine scientific community, was organizing and carrying out an attack on the image of the oceans as pristine and indestructible and was effectively formulating new norms for ocean protection.[7] While ocean pollution commanded the attention of the president and the administration, this group of congressmen saw the opportunity to minimize and perhaps even end ocean dumping, and to initiate a new oceans program.[8] From the point of view of the U.S. marine scientific community, the past decade had been dominated by "ocean rhetoric," but the newly established National Oceanic and Atmospheric Administration was welcomed as an opportunity to "get going."[9] Heightened public and political concern for the health of the oceans furnished them with the necessary window of opportunity. In order to mobilize public and political support for regulation, prominent environmentalists and experts were invited to congressional hearings, at which they described ecological threats to and even crises in the marine environment caused by pollution. Prominent ecologists and scientists attacked the view that the oceans have a capacity to absorb unlimited waste without harm to them.

This group of U.S. politicians saw the problem of ocean dumping as being "global in scope" (Sen. Jennings Randolph, quoted in *Congressional Record: Senate*, April 1, 1971, p. 9184).[10] National efforts alone would be futile. "We are faced not with a national problem," one politician declared, "but an international one. Unless the nations concerned combine to put an end to ocean abuse, the abuse will write finis to us all."[11] A Senate hearing highlighting the international character of ocean dumping was held in the fall of 1971. The goal of this hearing—officially named International Conference on Ocean Pollution—was to focus national and international public and political attention on ocean dumping and to demonstrate the need for international cooperation. The idea to convene this international conference was put forward in December of 1970 by a congressman who questioned the usefulness of the course of action suggested by the Council on Environmental Quality. He and other congressmen doubted the value of the Stockholm conference and routine diplomatic channels with respect to reaching an international agreement on ocean dumping. Accordingly, in a letter to President Nixon, one of the co-sponsors to this conference wrote: "I believe [the ocean dumping problem] is of such momentous importance as to warrant an international conference at which it could receive maximum attention. It seems to me that the exclusive attention which such an international conference could afford would be more productive of positive results than would be the case if we relied on a general conference such as the United Nation's Conference on the Human Environment scheduled for 1972."[12] Agreement reached at an international conference would, at the same time, be an important part of the United States' preparations for the Stockholm conference. As one senator put it, "the United States must be prepared to offer for consideration an international policy governing ocean disposal of materials." (Sen. Jennings Randolph, *Congressional Record: Senate*, April 1, 1971, p. 9184)

Representatives of the international diplomatic community attended the International Conference on Ocean Pollution.[13] Scientific aspects of ocean pollution were covered by Thor Heyerdahl, Jacques Cousteau, and Barry Commoner, all vocal international environmentalists and respected scientists. None of the marine scientists from earlier hearings participated.[14] Congress had been informed that in 1960 Cousteau had led a successful campaign to prevent the French Atomic Energy Commission from dumping radioactive waste into the Mediterranean.[15] While crossing the Atlantic in a reed boat in 1970, the Norwegian explorer Heyerdahl had collected samples of oil pollution, which later were displayed at United Nations headquarters in New York. Reports on oil pollution by Heyerdahl were also included in documents of GESAMP (the IMO/FAO/UNESCO/WMO/WHO/IAEA/UN/UNEP Joint

Group of Experts on the Scientific Aspects on Marine Pollution) and in background documents for the Intergovernmental Working Group on Marine Pollution (the negotiating and drafting group on the global ocean dumping convention).[16]

Senator Ernest ("Fritz") Hollings, chairing the International Conference on Ocean Pollution, explained the goal of focusing public and political attention on ocean pollution in his opening remarks: "This Second International Conference on Ocean Pollution is dedicated to putting people on the alert. Everyone talks a lot about ecology.... But we lack a sense of environment priorities.... We need much more a full-scale assault on the heart of the problem."[17] Barry Commoner's left-wing political views later almost overshadowed his scientific statement, and were met with strong objections by one senator.[18] At that point, Senator Hollings defined the crucial role of vocal ecologists and environmentalists in giving the ocean pollution issue the needed national and international visibility. "Specifically we all know," he said, "that the oceans program is dragging its feet. It dragged its feet under President Kennedy. It dragged its feet under President Johnson. It was due to this Congress that we got the Stratton Commission and President Nixon instituted the National Oceanic and Atmospheric Administration. We had a conference last week on how we could get the Administration going again in giving attention to the oceans, giving attention to the pollution problem, as the president gave in his Reorganization Plan No. 4 setting up the National Oceanic and Atmospheric Administration. So we are trying to move it along, and you have helped us in a magnificent way."[19] In his closing remarks to the international conference on ocean dumping, Hollings further emphasized the importance of the participation of Commoner, Heyerdahl, and the television personality Hugh Downs: "So the only way I know—I could say these things over and over again—but the only way we are ever going to get this message through is with people with the brilliance and dynamism of you three here this morning getting the attention of the American public and in turn of our colleagues here in the Congress to move in the right direction."[20]

Evidently the international conference's goal of focusing public and media attention on ocean pollution was achieved. Heyerdahl's statement was reprinted in the *Congressional Record*.[21] Cousteau's statement reached a much broader audience. On November 14, 1971, it was reprinted in an even more apocalyptic version in the *New York Times*,[22] under the headline

"Our Oceans Are Dying" (which has often been quoted in popular science publications and in the ecology literature[23]). On this occasion and on others, Cousteau significantly influenced public and political opinion with his message that "the oceans are dying" (interview, Robert J. McManus, August, 29, 1991).[24]

The View of Congress

The U.S. House of Representatives Committee on Merchant Marine and Fisheries released its report on ocean dumping in the summer of 1971. As to whether the oceans should be used for waste disposal, it said this: ". . . it seems fair to say that the Committee wished to emphasize its answer to that question as a very large 'No.'"[25]

The committee's report did not reflect the wide divergence of views among the scientists who had testified. Using carefully worded language, it noted "almost complete current unanimity of concern for the protection of the oceans from man's depredations." It continued: "In the hearings before this Committee, the witnesses were unanimous in their support for the purposes of this legislation. No argument was raised by any witness as to the desirability of creating a system of protection from unregulated dumping of waste material into the oceans."[26] As described in the previous chapter, there had been considerable disagreements among scientists and professional witnesses from the waste management field. But the committee chose to ignore the view of the latter group. To explain and justify the need for regulation of ocean dumping, the House committee quoted the prominent environmentalists Paul and Anna Ehrlich, Jacques Cousteau, and Thor Heyerdahl extensively, though only Heyerdahl had testified in the spring hearings.[27] An extensive quotation from Paul and Anna Ehrlich stressed nations' responsibilities toward one another: "No one knows how long we can continue to pollute the seas with chlorinated hydrocarbon insecticides, polychlorinated biphenyls, and hundreds of thousands of other pollutants without bringing on a world-wide ecological disaster. Subtle changes may already have started a chain reaction in that direction. The true costs of our environmental destruction have never been subjected to proper accounting. The credits are localized and easily demonstrated by the beneficiaries, but the debits are widely dispersed and are borne by the entire population through the disintegration of physical and mental health, and, even more importantly, by the potentially lethal destruction of ecological systems. Despite social, economic, and political barriers to proper ecological accounting, it is urgent and imperative for human society to get the books in order."[28]

During the spring hearings, covered in chapter 4, one congressman had submitted a letter from Jacques Cousteau, who was, in the words of the congressman, "the person most expert on the oceans of the world" and the individual whose "testimony is the best available to alert us to the damages we have done to our oceans and to the dangers we face if we do not act quickly and constructively."[29] The House report quoted a portion of Cousteau's letter: "Because 96 percent of the water on earth is in the ocean, we have deluded ourselves into thinking of the seas as enormous and indestructible. We have not considered that earth is a closed system. Once destroyed, the oceans can never be replaced. We are obliged now to face the fact that by using it as a universal sewer, we are severely over-taxing the ocean's powers of self-purification. The sea is the source of all life. If the sea did not exist, man would not exist. The sea is fragile and in danger. We must love and protect it if we hope to continue to exist ourselves."[30] The report also underscored the global scope of the ocean dumping problem. Heyerdahl had "found evidence of pollution and dumping of materials throughout his trip from Africa to the West Indies," and "these issues formed the focus and background for the hearings on the Administration's ocean dumping legislation."[31]

The administration's proposal to ban the dumping of chemical and biological warfare agents and high-level radioactive waste had the full support of the committee. The committee considered high-level ("hot") radioactive waste so hazardous that it recommended an absolute ban on disposal at sea. A spokesman for the Atomic Energy Commission had assured the committee that the AEC did not consider resuming ocean dumping of low-level waste (which, as described in chapter 2, had been almost completely phased out since 1963 in the United States).[32] Hearing reports recommended that an international agreement on the dumping of radioactive waste at sea be established. They included papers, provided by conservation groups, advocating "restraint and careful planning in nuclear exploitation of the oceans" and stressing an urgent need for "worldwide agreements limiting radioactive pollution."[33]

The more detailed report of the Senate Committee on Commerce covered both pro-dumping and anti-dumping views. Nonetheless, it was unmistakably precautionary and prohibitory in tone. It quoted extensively from Cousteau's statement at the International Conference on Ocean Pollution (held after the House report was issued), and it frequently used images, arguments, and passages from that statement. A portion of the report said: "We have treated the oceans as enormous and indestructible—145 million square miles of surface—the universal sewer of mankind. Previously we thought that the legendary immensity of the ocean was such that man could do nothing against such a gigantic force. But the real volume of the ocean is very small when compared to the volume of the earth and to the volume of toxic wastes that man can produce with his technological capability. The water reserve on our spaceship is very small. And again, as Captain Cousteau has said: 'The cycle of life is intricately tied with the cycle of water. Anything done against the water is a crime against life. The water system has to remain alive if we are to remain alive on this earth.'" (Marine Protection, Research, and Sanctuaries Act of 1972, Senate Report 92-451, 92nd Congress, 1st session, 1971, p. 4237) Parts of this passage, which emphasized norms and ethical concerns but which contained little scientific evidence, were later quoted repeatedly in the Senate debate on the dumping bill.[34]

The Council on Environmental Quality and the Department of State pointed out during the hearings that apparently all waste dumped off U.S. coasts originated in the U.S., and that this situation was not expected to change in the future (Russell Train, "Statement," in U.S. House of Representatives, Ocean Dumping of Waste Materials, p. 170; John Stevenson, "Statement," in U. S. Senate, Ocean Waste Disposal, p. 193). Foreign dumpers did therefore not represent a threat to U.S. waters. From a rational economic point of view, there was no evidence that a global regime was needed to compliment U.S. domestic action.[35] The situation was different in Europe, however, where geographical proximity and shallow seas created a disincentive for unilateral action.

Instead of reasoning as realism's rational egoists exclusively concerned with the protection of American national interests, important U.S. politicians thus perceived ocean dumping as a fundamentally global environmental problem concerning all humankind. The House committee's report said: "The Committee wishes to emphasize its awareness that the types of problems with which [the ocean dumping bill] deals are global in nature. We are not so blind as to assume that in dealing with the problems created by our own ocean dumping activities, we are thereby assuring the protection of the world's oceans for all mankind. Other nations, already moving to grapple with these troublesome issues, also will and must play vital roles in this regard." (House Report no. 361, p. 14) A sense of guilt for past polluting activity was a further incentive for the United States to take the initiative to begin controlling ocean dumping globally: "The committee recognizes that the United States has been heavily involved in ocean dumping activities and that the kinds of materials that our highly industrialized, commercial nation may be forced to dispose of may be particularly hazardous to the health of the oceans." (ibid.) Furthermore, in terms more commonly used by ideational scholars and reflectivists than by realists and power theorists, the House committee's report urged the United States to play a leadership role: "Importantly, we believe strongly that someone must take the first steps." (ibid.) The need for U.S. leadership had also been stressed at the earlier hearings.[36]

In Congress, Cousteau's statements (originally published in the *Washington Post* and reprinted in the *Congressional Record*) that "the oceans are in danger of dying" and "the pollution is general" were repeated frequently, as were his and Heyerdahl's descriptions of pollution encountered in isolated and previously unspoiled parts of the oceans.[37] Cousteau's earlier support for the Nixon administration's bill was also repeated. The soundness of the two explorers' statements and policy advice was not questioned. Debatable knowledge was not debated. Moreover, the abundant scientific uncertainties that surrounded ocean dumping necessitated immediate action instead of restraint. Senator Hollings declared: "The seas are dying according to Jacques Cousteau, but we have not done much to find out whether he is right or not. And if we wait much longer, we may not have the luxury of time to find out. Because if the oceans die, we die." (Sen. E. Hollings, *Congressional Record: Senate*, November 24, 1971, p. 43074) Some proposed even more stringent regulation.[38]

It was, as would be expected, congressmen from states contiguous to the Great Lakes and from the coastal states who most actively supported control of ocean dumping. Undoubtedly, politicians also felt pressure to demonstrate willingness to protect the environment.[39] But the high number of votes with which the bill passed both in the House and the Senate reflected genuine concern over the environment among the public as well as political and administrative leaders.[40] One Republican senator saw the United States' ocean dumping regulation as "the result of our relatively sudden realization that the sea is not a bottomless septic tank, but a delicately balanced ecosystem dependent upon the good sense of man for its continued existence" ("Dumping of Waste Materials," *Congressional Quarterly: Weekly Report*, December 11, 1971, pp. 2548–2549). Similarly, on Congress's motives for overwhelmingly supporting an amendment to a water pollution bill to study limitations on DDT, the *Congressional Quarterly* reported in the spring of 1970: "To a certain extent, Members were rushing to get on the bandwagon. 1970 is an election year, and public concern over the fate of the environment has never been higher. But evidence indicated that much of the new concern in Congress is sincere—and may continue after the current frenzy of activity slows down." ("Pollution: Everyone Wants a Piece of the Action," *Congressional Quarterly: Weekly Report*, April 24, 1970, p. 1135) Several conservation and wildlife preservation groups supported ocean dumping regulation but were not actively involved in the political process.[41] Thus, leadership by influential politicians and ecologists, together with pro-environment mass media, was crucial, as public concern over ocean dumping and marine pollution was only moderate despite attempts to focus attention on these particular issues (interview, Charles Lettow, September 24, 1991).[42]

Economic Costs of U.S. Ocean Dumping Regulation

The economic costs of U.S. domestic ocean dumping regulation gave rise to concern. Like other pieces of regulation to protect the environment, domestic regulation would impose economic costs on U.S. industry which, in the absence of global regulation, would benefit foreign industry. In the case of ocean dumping, U.S. government representatives and industrialists were concerned about the possible economic implications regarding European and (to a lesser extent) Japanese industries, and great importance was attached to finding an international forum where those problems also could be tackled.[43] Thus, the importance of institutional choice in a new era of policy making was evident.

Because Japan was not a member of the North Atlantic Treaty Organization, the administration disapproved of using NATO as the primary forum for negotiating an international agreement on ocean dumping.[44] Because its member countries were quite similar relative to the countries participating in the Stockholm conference, the Organization for Economic Cooperation and Development was favored by the United States for dealing with international trade implications of environmental regulation.[45] The exclusion of the Soviet Union from the OECD was seen as the major drawback of that organization.[46] In the case of a global ocean dumping convention and other problems of the marine environment, therefore, the United States preferred to work through the United Nations system, in particular through the Stockholm conference, as it included nations with substantial oceanographic capabilities (e.g. the Soviet Union, Japan) and large maritime fleets (e.g. Liberia, Panama).[47]

While the economic implications of ocean dumping regulation were modest, but still gave rise to concern, the potential economic costs of international environmental regulation that might be agreed on in Stockholm caused serious concern in the United States. The economic consequences of differing national standards jeopardized international trade. Furthermore, an international trade war seemed a real threat as there was considerable pressure within Congress and the Nixon administration to impose countervailing duties where other countries did not maintain standards comparable to U.S. standards. The U.S. Assistant Secretary of Commerce warned of the possibility of a trade war in a speech on October 6, 1970: "In those cases where prices increase (to meet pollution control costs), U.S. goods would be at a competitive disadvantage in world trade. In order to avoid a major deterioration of our balance of payments position, remedial action would be necessary. Perhaps the most desirable action would be the setting of international pollution standards. An international convention of the world's countries could be convened for the purpose of reaching agreement on pollution standards. If an international agreement on pollution standards cannot be reached, the U.S. may find it necessary to levy border taxes on imports and rebates on exports to reflect the added production costs of pollution standards. This is obviously a less desirable solution, because it might violate existing GATT [General Agreement on Tariffs and Trade] rules and because it would be difficult to determine the extent to which the imposition of pollution standards adds to production costs."[48] Such a trade war would probably escalate as the environment increasingly became an issue also in Europe; the environment was already an issue in Japan. The U.S. Department of State thus intended to use the Stockholm conference for reaching agreement on international regulations and standards in order to protect national economic interests and avoid trade disruptions.[49] Similarly, to protect the United States' economic interests, some senators wanted the U.S. delegates to the Stockholm conference to "advocate and support multilateral accords . . . enforceable by the United Nations or multilateral economic sanctions."[50]

Also, in the case of ocean dumping, international agreement would offer an international solution to the economic costs imposed by domestic regulation. A U.S. official participating in the regime negotiations explained: "One reason the United States strongly supported an ocean dumping treaty in the first place was its hope that other nations—especially industrialized ones—will establish environmentally protective regulations similar to our own. To the extent they do not, foreign industry may gain a competitive edge, since the price of its products will not reflect the costs of pollution abatement. And so, once enactment of domestic ocean dumping legislation was foreseen, the United States became enthusiastically instrumental in establishing an international control mechanism reflecting our domestic law." (McManus 1973, p. 26)[51] In short, agreement on international regulation would imply that also foreign industry should reflect the costs of pollution control in its products.

Thus, on the economic side, soon-to-be realized domestic regulation created incentives to establish a global environmental regime and prompted U.S. leadership in ocean dumping regulation. To protect national economic interests, the United States was in strong support of international regulation as this could provide an acceptable solution to the economic costs of domestic regulation. The Stockholm conference was the preferred forum for reaching agreement on international regulation, both from the economic and the environmental perspective. Hence, the United States tabled a proposal for a global ocean dumping convention at the first meeting of the Intergovernmental Working Group on Marine

Pollution, a negotiation group established by the Preparatory Committee of the Stockholm conference.

Conclusions

In the late 1960s, several spectacular mishaps and incidents of pollution had attracted American attention to pollution of rivers, lakes, and harbors. Attention to pollution of the ocean from dumping, however, was provoked by the problem of dumping dredged spoils into the Great Lakes and by the U.S. Army's dumping of nerve gas off the coast of Florida. Although no severe damage had been inflicted on the ocean environment, ocean dumping was singled out for regulation. This approach of (in President Nixon's words) "acting rather than reacting" to prevent marine pollution from dumping also made political sense to politicians who were under pressure to demonstrate willingness to act against conspicuous pollution threats.

Public concern about the health of the oceans was new in the late 1960s. It was a fortunate combination of a series of focusing events, a new influential public idea, and a group of determined legislators and environmentalists acting as policy entrepreneurs that resulted in regulation of ocean dumping in the United States. This combination of elements had been mentioned in the opening remarks at one of the congressional hearings on ocean dumping in 1971: "It seems that no one knows the volume—and I think that is really an understatement—of wastes that have been dumped in the oceans in the past years. In fact, until recently, the question was scarcely asked and then only by an obscure group of scientists, known as ecologists. Fortunately, however, in the last few years the entire question of ocean disposal of waste material has been thrust into prominence, and I think appropriately so, by the recently disclosed dumping of nerve gas and oil wastes off the coast of Florida, by the dumping of sewage and other municipal wastes off New York Harbor, and a number of other and similar instances."[52]

In the wake of a series of spectacular environmental accidents in the ocean and examples of dumping causing "death" in Lake Erie and other large water bodies, the idea of pollution of the ocean from dumping, which had started among an insignificant group of scientists, was used by a group of congressmen and prominent international ecologists and environmentalists to effectively set new norms for ocean protection. To attract public and political attention to the need for regulation, this transnational coalition of policy entrepreneurs spread the simplistic, powerful idea that the oceans were "dying."

Contradicting the claim made by realists, the hegemon perceived ocean dumping as a global environmental problem involving all states. Protection of the national and collective interest required that all states work together in controlling ocean dumping. But an international policy would be realized only if states could agree on a definition of the problem that needed to be solved. This could not be achieved by coercion, co-optation, and the manipulation of incentives; it called for international persuasion and education. Though chapter 6 will focus on those aspects and stages that the interest-based regime approach addresses in most detail, it will also demonstrate that realist propositions do not explain the establishment of the global ocean dumping regime satisfactorily.

6
Negotiating the Global Ocean Dumping Regime: Interest-Based Regime Analysis

For neoliberals, global environmental problems make states dependent upon one another for the attainment of human well-being and protection of the environment, and this interdependence almost compels states to cooperate within regimes. According to this body of theory, military force is generally not an available means to influence international policy coordination under conditions of complex interdependence.

Many neoliberals would expect that the United States—or, alternatively, a group of powerful states—would construct the global ocean dumping regime, with international organizations playing an insignificant role in the process. According to a classic neoliberal study, "leadership will not come from international organizations, nor will effective power" (Keohane and Nye 1977, p. 240). Other neoliberals would expect entrepreneurial leaders to play a prominent role in negotiations producing environmental regimes, whereas structural and (especially) intellectual leaders probably would play less significant roles.

The formation of the global ocean dumping regime was far from a "governments only" affair. It was intimately interwoven with the 1972 Stockholm conference, where organizers attempted to secure the environment a permanent place on the global agenda. As will be documented in this chapter, the Stockholm secretariat organized the negotiations on the global ocean dumping regime so that a treaty could be ready for signature by governments at the Stockholm conference. It was believed that agreement on this global environmental regime, which was in fact reached within a few months after the Stockholm conference, would prove governments' willingness to start protecting the environment.

It will become evident that interest-based analysis captures many but not all of the elements of how, why, and by whom the global ocean dumping regime was built. This chapter demonstrates that policy entrepreneurs inside the Stockholm secretariat played an important catalytic and facilitative role in the pre-negotiation and negotiation stages of regime formation. The secretariat was actively involved in framing and communicating ocean pollution as an urgent global policy problem. Serious tensions between developed and developing countries, major differences in states' commitment to protecting the environment, and lack of global attention to ocean dumping jeopardized the construction of the regime. Differences in domestic policies and national positions and their possible international implications clearly manifested themselves over the course of the negotiations. However, largely through the policy entrepreneurship of the Stockholm secretariat, these obstacles were overcome. As this chapter will also document, international pressure was a significant element in reaching agreement on the regime. As chapter 7 will demonstrate, the secretariat effectively mobilized international public opinion and pressure during the preparations for the Stockholm conference.

The Stockholm Conference and Ocean Pollution from Dumping

The idea to convene a high-level United Nations conference in order to focus the attention of the international community on the need for international action on the environment originated with Sverker Aström, head of Sweden's mission to the United Nations. A member of the Swedish delegation made the formal proposal at a meeting of the UN's Economic and Social Council in July of 1968. The council's resolution calling for the conference was then debated by the UN General Assembly. The assembly adopted the draft document without alteration in December of 1968. Under the resolution, the coming United Nations conference was to "provide a framework for comprehensive consideration within the United Nations of the problems of the human environment in order to focus the attention of Governments and public opinion on the importance and urgency of this question and also to identify those aspects of it that can only, or at best be solved through international cooperation and agreement" (UN General Assembly Resolution 2398 (XXIII), December 3, 1968, quoted in Caldwell 1984, p. 44).

Michel Batisse, the organizer of the UNESCO Biosphere Conference in 1968, appointed a Swiss scientist, Jean Mussard, as Secretary-General of the coming conference. Mussard planned to organize an international meeting that would focus on scientific aspects of the environment. Political and economic aspects of environmental protection were ignored. At that point, the coming conference seemed more likely to produce scientific reports and books, and probably financial support to UNESCO, than to produce concerted governmental action.

The Food and Agriculture Organization and the World Health Organization, both bypassed in the early planning phase, did not share UNESCO's intentions for the coming conference. Sverker Aström, who said in his published recollections that "from the very beginning we emphasized the need for rapid action" (quoted in Rowland 1973, p. 34), realized that the preparations did not develop as he intended. Aström and a few high-level United Nations officials persuaded U Thant, then Secretary-General of the UN, to replace Mussard. In December of 1969, a UN resolution shifted the direction of the coming conference "to serve as a practical means to encourage and to provide guidelines for action by governments and international organizations" (General Assembly Resolution 2581 (XXIV), December 15, 1969, quoted in Rowland 1973, p. 35). In the autumn of 1970, Maurice Strong, a former businessman who had recently been appointed as head of the Canadian International Development Agency, replaced Mussard as Secretary-General of the conference.[1] An international policy conference then began to take shape. The official title of the conference was changed, as both the FAO and the WHO had hoped, to emphasize the "human aspects" of the environment.[2] This shift implied that political and economic aspects, as well as the proper role of science and technology in environmental protection, should move to the fore. An action-oriented approach to the preparations for the conference as well as for the conference itself was also developed by the Stockholm secretariat.

The members of the secretariat (numbering about 20) were aware that a single United Nations conference on the environment could not suddenly bring governments to massively cooperate on environmental protection. Many developing countries suspected that environmental protection was simply another way for developed countries to slow down their industrial development. At the same time, protection of their economies and sovereignty made developed countries oppose any bold attempts at international cooperation on these matters. The secretariat nonetheless hoped for the beginning of environmental protection on a global scale.

The secretariat established a Preparatory Committee, consisting of 27 governments, with strong representation from the Third World, which at its first two meetings looked for particular areas for future international cooperation on pollution control. The UN General Assembly resolution to "identify those aspects of it that can only, or at best be solved through international cooperation and agreement" led the Preparatory Committee in their search for parts of environmental problems that were joint (interview, Peter S. Thacher, May 2, 1991). The committee focused on areas outside national authority and areas under national authority of concern to most governments.

The secretariat tried to convince governments that a number of pollutants and ways in which pollution occurred could, because of their nature, be solved only, or best, through international cooperation. Among these pollutants of "broad international significance" (as they were called in "basic papers" that were produced by specialized agencies of the United Nations and presented during the preparatory process), three types were identified: those whose effects were felt beyond the national jurisdictions in which the pollutants were released to the environment, those that affected international trade, and those that occurred in many states.[3] Global aspects of marine pollution were emphasized in separate basic papers.[4] The secretariat hoped to demonstrate that there were particular pollutants that, from a global perspective, should be controlled. The oceans, the stratosphere, and Antarctica seemed strong candidates for the first category, while maintenance and restoration of soils and conservation seemed strong candidates for the third category. However, protection of the marine environment attracted by far the most attention as an area where it was hoped action could be undertaken in Stockholm. The secretariat then took the initiative to convene an international working group on marine pollution, officially named the Intergovernmental Working Group on Marine Pollution (IWGMP), to prepare action in this particular field.

Not all governments saw ocean dumping as a global environmental problem, however. A number of governments were in fact not concerned about ocean dumping. Only in those countries where environmentalism had gained a foothold, namely developed countries and especially the United States, did ocean dumping cause concern. It was in the United States, where prophecies of environmental catastrophes were a peculiar characteristic of the ecology debate as well as a favorite theme of the ecology elite, that the "dying oceans" idea first gained political importance and ocean dumping was first seen as a global environmental problem. A few months before the

Stockholm conference, Hawkes (1972a, p. 738) reported: "The many environmental meetings that have preceded Stockholm have shown that Americans tend to take a much gloomier view of the situation than do Europeans. . . . The more apocalyptic visions of the future remain a minority taste in Britain."[5] The U.S. government had evidently taken up this idea. A former U.S. ambassador to the United Nations later noted caustically: "We have become great producers and distributors of crisis. The world environment crisis, the world population crisis, the world food crisis, are in the main American discoveries—or inventions—opinions differ." (Moynihan 1978, p. 131)

Especially in the eyes of China, Brazil, and India, the affluent Western societies, which were facing severe, self-inflicted environmental problems, intended to use the Stockholm conference to impose new environmental regulations on developing countries that would cause industrial and economic stagnation. Brazil's foreign minister argued at a "Group of 77" meeting that global pollution was "a by-product of the intensity of industrial activity in the highly developed countries" (Hawkes 1972a, p. 737). At a meeting preceding Stockholm, some developing countries even saw global environmental standards as a deliberate strategy of the developed countries aimed at halting the industrial and economic development of poor countries (ibid.). In Stockholm, developing countries responded by launching a verbal attack on the developed countries, especially the United States, and demanded compensation and assistance in development. According to American mass media, this was an "unexpected theme" (Hill 1972).

As regards ocean dumping, the view of the group of developed countries and that of the group of developing countries differed dramatically. Developing countries generally did not see themselves as polluters of any significance and did not consider ocean pollution their problem. Nor did scientists from developing countries pay much attention to ocean pollution. A spokesman for scientists from the developing countries had explained in a hearing before the U.S. Congress: "Ocean and higher atmosphere pollution, that is to say the two phenomena with the greatest global effects, have almost not been considered [by Third World scientists]. I would dare to interpret this fact as the feeling that the less developed countries are judging themselves only in a very small measure responsible for the occurrence of these pollutions and that therefore the solutions should also be undertaken by the industrialized countries."[6] In his comments on the Stockholm secretariat's proposal to identify and list pollutants of "broad international significance," the Brazilian delegate to the Stockholm conference observed that "the great polluters are the highly industrialized countries. Starting from radionuclides (practically 100 percent of whose production and dissemination is imputable to a few highly developed countries) and going right on down the list of all the other major pollutants, the overwhelming discharge of effluents is the consequence of the developed countries' recent technologies and of their high levels of industrial as well as primary production.... The contribution to this type of pollution by underdeveloped countries is, in absolute terms, extremely small and in relative terms practically nil." (de Almeida 1972, p. 48)[7] A 1971 report from the Indian National Science Academy likewise did not count ocean pollution among existing global environmental problems.[8] Despite these clear indications of opposition to global environmental standards and modest interest in marine pollution control, the U.S. delegation was surprised, as was mentioned in chapter 4, when it learned that developing countries preferred not to establish a global regime controlling ocean dumping.

But the Stockholm secretariat was not the least surprised by the developing countries' opposition to global environmental standards. At meetings arranged by the secretariat, representatives of developing countries had made it clear that they were opposed to such standards.[9] However, Maurice Strong assured in his round-the-world lobbying to organize the conference that land use, drastic erosion, spreading deserts, and loss of wetlands and watersheds—all topics most relevant to developing countries—would be on the Stockholm agenda.[10] In addition, conference organizers kept family planning—a politically sensitive issue—off the agenda in Stockholm, hoping not to alienate developing countries.[11] As the journal *Science* pointed out, this was "part of the price paid for persuading underdeveloped countries to come to a conference they have no heart for" (Hawkes 1972a, p. 737). A United Nations conference on population was instead planned for 1974.

The Stockholm Secretariat and Ocean Dumping

As has already been mentioned, the Stockholm secretariat took the initiative to establish the Preparatory Committee and later the IWGMP. As early as November of 1970, the secretariat singled out ocean dumping as a strong candidate for global agreement in Stockholm and advocated agreement in this area at meetings with the Preparatory Committee.[12] The secretariat was searching for problems which governments collectively could start addressing by establishing a global regime. Although land-based sources of marine pollution (rivers, pipelines, outfalls, and runoff) supposedly accounted for as much as 90 percent of marine pollution, insufficient knowledge about these sources and much higher economic costs associated with their control made land-based marine pollution a very unlikely candidate for international regulation. Scientific evidence of pollution from ocean dumping also made this issue easier to tackle. Since oil pollution from ships would be dealt with at an Intergovernmental Maritime Consultative Organization conference in 1973, dumping was left as the "last key maritime source" of ocean pollution. The secretariat thus saw the possibility to "close off one remaining source" and in this way demonstrate that governments were able and willing to act (interview, Peter S. Thacher, August 14, 1991).

To help negotiations on the global ocean dumping convention, the secretariat introduced the notion, innovated by lawyers and scientists within the United Nations, that substances could be classified into so-called black and gray lists by an international convention (interviews, Sachiko Kuwabara, August 26 and 27, 1991). According to this notion, dumping of blacklisted substances should be prohibited and special care should be taken and permission should be given before graylisted substances were dumped.[13] The secretariat recognized that categorizing substances on the basis of available knowledge of their environmental impact in separate lists annexed to the convention had several advantages. For one, the contents of the lists could be updated and adjusted as new knowledge of pollutants developed without it being necessary to negotiate the entire convention text. This alone would mean a major innovation of the standard treaty form.[14]

Recent experiences showed that lack of knowledge about pollutants was a significant barrier to international agreement.[15] The system of black and gray lists also offered a solution to this problem in that the lack of complete knowledge concerning the effects of pollutants on the marine environment could be acknowledged without jeopardizing agreement. As one U.S. delegate to the negotiations later described, this approach helped significantly in overcoming this particular obstacle to agreement: "There was very little disagreement over the scientific portions of the Convention. All accepted the concept of an annexed list of substances banned from ocean dumping. Another annexed list would contain substances requiring special care before dumping could be permitted. The delegations recognized that present knowledge of the effect of substances in the marine environment was quite deficient, and accordingly, the lists were prepared in light of a rapid amendment procedure for the annexes." (Lettow 1974, p. 665)

Furthermore, the secretariat was aware that using black and gray lists was a moderate, piecemeal approach that allowed reluctant governments to join a global ocean dumping regime and thereby be looked upon as pro-environment at a time where the public, especially in the Western world, was considerably concerned about the environment. This would also make good political sense for developing countries, as they feared that future "environmental aid" would reduce existing funds for development aid and that public pressure in donor countries would channel resources away from developing countries looked upon as being "anti-environmental." Since the system of black and gray lists effectively gave governments the option of not doing everything immediately, the secretariat hoped that it would keep an all-or-nothing dilemma from arising in the negotiations. It was also hoped that the approach would help delegations to overcome resistance to joining a global regime home in their capitals. One member of the Stockholm secretariat summarized this particular advantage of the lists as follows: "The black list allows you to have a gray list." (interviews, Sachiko Kuwabara, August 26 and 27, 1991)

As has already been pointed out, the secretariat realized early on that developing countries were very skeptical about the need for international regulations and standards. Thus, it was considered very important that developing countries not suspect that the United Nations was on the side of the developed countries, and the UN wished to demonstrate the existence of global interest in controlling certain pollutants. It was hoped that the "basic papers" on pollutants of "broad international significance" would help to accomplish this. The secretariat hoped in particular to appease the fear of some developing countries, most importantly Brazil, toward agreeing to control at least some pollutants.[16] United Nations experts had furthermore recognized that it was very unlikely that governments would support the creation of an international agency with planning and enforcement powers.[17] They hoped instead for "direct cooperation between non-diplomatic officials in different countries" (Schachter and Serwer 1971, p.

104). They also hoped to separate the technical from the diplomatic—science from politics—in international environmental negotiations, as the black and gray lists would.

The secretariat was aware that the use of black and gray lists had other advantages that might have constructive influence on the negotiations. Primarily, the approach was instrumental in reaching agreement on banning dumping of at least some pollutants. Further, it demonstrated to developing countries that developed countries were serious about controlling ocean dumping.[18] Developing countries generally would have no reason to worry about the economic consequences of ocean dumping regulation, since they would have few if any of their substances blacklisted. Thus, in reality the black and gray lists imposed heavier burdens on those governments who were more concerned over pollution, and lesser burdens on those who were not as concerned. As detailed in this chapter, developing countries favored stringent black and gray lists but were opposed to draft general provisions which they feared would impose unacceptable constraints on their economies. In conclusion, the Stockholm secretariat promoted the relatively straightforward notion of black and gray lists because it had important advantages in environmental negotiations with participation of developed and developing countries.

Negotiating the Global Ocean Dumping Regime

Within a year, the Intergovernmental Working Group on Marine Pollution met four times. At the fifth session, held in London from October 30 to November 13, 1972, a global dumping convention was signed. This group, established by the Stockholm secretariat as part of the preparations for the Stockholm conference, was intended to produce an action program for future international control of marine pollution to be presented at the conference.

The First London Session
The first session of the IWGMP took place on June 14–18, 1971, at IMCO headquarters in London, under the sponsorship of the British government. Thirty-three states had sent representatives.[19] Maurice Strong and representatives of the Stockholm secretariat, GESAMP, the International Oceanographic Commission, the FAO, the IAEA, the WHO, the WMO, and

UNESCO also attended, as did representatives of IMCO. An observer from the UN's Group of Experts on Long-Term Scientific Policy and Planning (GELTSPAP) was also present. Thor Heyerdahl, who participated as a special adviser to the Norwegian delegation, described having encountered "visible signs of extensive pollution" on his two transatlantic voyages.[20]

After a brief speech by the British environmental secretary, Maurice Strong addressed the IWGMP. He first underlined the common interests of states in cooperation on marine pollution control: "We are dealing with the problem of more than seventy percent of the surface of this planet; seventy percent of this planet's biosphere on which all life depends—and most of this beyond the protection of any nation or any group of nations."[21]

Strong was at pains to stress that sufficient knowledge existed to act against specific pollutants, and he underscored that international action was urgently needed and, despite many unknowns, scientifically justified: "It must be acknowledged that we lack sufficient knowledge today on which to base *all* future decisions. . . . But we do know enough to begin to take some of the important decisions that must be made. . . . The recommendations of GESAMP are clear on this point; that the time for action is at hand." (Annex IV to A/CONF.48/IWGMP.I/5, p. 2) In the face of incomplete scientific knowledge, a piecemeal approach would facilitate cooperation in certain areas. At the second session of the IWGMP, Strong similarly insisted that governments needed "no longer await the results of painstaking scientific research; they already knew enough to act."[22]

Strong vehemently advocated precaution and immediate action. He cautioned that economic assessment likely would underrepresent damages from ocean dumping and moreover delay necessary action: "[Economic] figures would represent only the tip of the iceberg of ultimate costs. Just as the law must anticipate science to a certain extent [because of long-term effects of pollutants], I think we will agree that the law cannot wait either on a fully detailed cost-benefit balance sheet."[23] Concluding, he repeated the need for action by governments: "So we have now reached the point where we need to get down to work if we are to prepare the concrete proposals which are so urgently needed for action at Stockholm. . . . This Group could demonstrate that we have moved from the talking stage to the action stage. This is, I believe, what the world expects of us." (ibid.)

From the outset, the participants agreed that many forms of action were needed, owing to the variety of human activities on land and at sea, to prevent marine pollution. Yet, in addition to drawing up a comprehensive plan for preservation of the marine environment, it was decided to single out particular aspects of marine pollution for which it might be possible to conclude a treaty at the Stockholm conference (A/CONF.48/IWGMP.I/5, p. 5). A U.S. draft convention ("Regulation of Transportation for Ocean Dumping Convention") and regional arrangements to protect seas or groups of seas were the main topics mentioned in this connection. Although discussions on regional arrangements would have to wait until the next session, since detailed proposals did not exist, widespread support for a global arrangement for dumping control was expressed.[24]

Under the U.S. draft convention, the convention would apply to dumping by all means of transportation; other sources of marine pollution (for instance, land-based sources, such as pipelining from the coast) would be excluded from the convention. Transportation of all materials from land for the purpose of dumping at sea would be prohibited unless a permit was issued by relevant state authorities. For that purpose, each state should establish criteria for the issuance of dumping permits. These criteria, which gave rise to debate, generally should be designed to "avoid degrading or endangering human health, jeopardizing marine life, and economic uses of the ocean" (A/CONF.48/IWGMP.I/5, p. 6). States should then notify an international registry as to the kinds and amounts disposed of, the location of the disposal site, and other relevant data.

The question of the prohibitory and restrictive stance (or lack of same) of the U.S. draft convention stirred debate. A number of states believed that the convention should primarily prohibit ocean dumping and should consider for disposal only those materials "whose harmless effects could be demonstrated in the light of existing knowledge and experience" (ibid., p. 6). Referring to the work done by GESAMP, it was suggested that whereas dumping of some substances might be prohibited, dumping of others might be allowed, subject to a license. Other countries found the suggestions of the U.S. draft more acceptable.

The U.S. draft did not distinguish between permitting and prohibiting ocean dumping.[25] Sweden, which in January of 1972 passed a law prohibiting ocean dumping altogether, found it too lax. Moreover, the criteria for issuance of dumping permits left too much discretion to individual states. Canada suggested that elements of the GESAMP definition of marine pollution be incorporated in the U.S. proposal. Since the draft convention did not specify what criteria should be followed to distinguish between a general permit and a special permit, it was also inquired what the limits of the general permit would be. Speaking in reply, the United States agreed, to some extent, with the views expressed, and declared that the draft would be revised to take into account the points raised with respect to dumping criteria and the need for international principles and guidelines to harmonize national regulatory approaches. In general, the draft was intended only as a first step toward an international regulatory arrangement.

With respect to pollutants to be regulated by the convention, it was agreed to focus on a number of pollutants: urban effluents; oil; toxic and persistent substances (e.g., organochlorine pesticides and other persistent chlorinated hydrocarbons); metals that accumulated in the food chain (e.g., mercury and other heavy metals); and industrially produced organic wastes (e.g., pulp and paper mill wastes and organic wastes from refineries) (A/CONF.48/IWGMP.I/5, p. 8). The list was not intended to be definitive, and it could be revised to take into account scientific advice given by advisory bodies such as GESAMP. In fact, the listed pollutants were identified on the basis of advice given by GESAMP. Importantly, radioactive substances were not among the listed pollutants. Pointing out that existing arrangements already were in place (see chapter 2), the United States and Britain did not intend to include radioactive wastes in the list of pollutants.

As the first meeting came to the end, a general optimism prevailed as to the possibility of reconciling the different approaches to regulation of ocean dumping. However, whether the primary emphasis should be on prohibiting ocean dumping remained a question. Canada and some other countries favored a strong anti-dumping policy, whereas the U.S. draft convention made no mention of prohibition of dumping.[26] In the view of the Canadian delegation, the U.S. draft proposed little more than a "license to dump" regime (Duncan 1974, p. 300). Nonetheless, the IWGMP decided to continue negotiating a global dumping convention. A feeling of urgency was unmistakable. In the unusually forthright and impelling words of the meeting report's summary of conclusions: "There are specific actions which should be prepared for completion at the time of [the Stockholm conference]. Although recognizing that final answers for many serious marine pollution problems must await more complete understanding of the marine environment, certain particular actions were identified which, if taken in the near future, could materially improve the situation and serve as evidence of the utility of international co-operation to protect the oceans. Proposals with regard to the control of ocean dumping are a specific step for which preparation should go forward with urgency." (A/CONF.48/IWGMP.I/5, p. 13) A global ocean dumping convention appeared to be within reach in the immediate future.

The Ottawa Session
The second session of the IWGMP took place in Ottawa on November 8–12, 1971. It was attended by representatives from 41 states and representatives from the Stockholm secretariat, GESAMP, the FAO, UNESCO and its IOC, the WHO, UNITAR, the WMO, IMCO, and the IAEA.[27] In general discussion, the IWGMP "reaffirmed the importance of urgent and effective action against marine pollution, especially by dumping" ("Report of the Intergovernmental Working Group on Marine Pollution on Its Second Session,"A/CONF.48/IWGMP.II/5 (November 22, 1971), p. 7)

Although recent progress toward regional institution building was welcomed by the IWGMP (the Oslo Convention, officially the Convention for the Prevention of Marine Pollution by Dumping from Ships and Aircraft, had been drafted October 22, 1971), many states were in agreement that action at a global level was necessary in order to link together and complement regional arrangements. Moreover, several developing countries thought that the Oslo Convention should not serve as a model for the global dumping convention. Instead, it was crucial that a convention not have loopholes that would allow developed countries to dump substances that under no circumstances should be dumped. This point was repeated during the final negotiations at the London Conference, particularly by the developing countries (memo, Ministry of Foreign Affairs, Denmark, December 6, 1972, p. 3). But a global convention should, on the other hand, not hinder the industrialization of developing countries. Brazil had made this clear at the first session.[28] In addition, some of the smaller Western European countries, including the Netherlands and Finland, were concerned about the risk of excessively vague rules of exemption from the lists of substances banned by a global convention (memo, Ministry of Foreign Affairs, Denmark, December 6, 1972, p. 3). To that end, a drafting group set up on an open-ended basis produced a number of provisional articles.

Several draft articles took a firmer prohibitory and restrictive stance than the first U.S. draft convention.[29] Moreover, although not yet specified, a distinction between "general permits" and "special permits" was made.[30] In addition, dumping of a number of substances was directly prohibited. Importantly, disagreement as to whether radioactive waste should be regulated by the convention had emerged. Draft articles put mentions of radioactive wastes in brackets, indicating that the issue was unresolved.[31]

"North-South" conflicts did not significantly influence the Ottawa session. Anticipating the coming conflict, Spain recognized among several identified "duties of international cooperation" the need for assistance from "states at higher levels of technological and scientific development" to states that would request it (A/CONF.48/IWGMP.II/5, Annex 4, paragraph 11). Time did not permit discussion of this principle.

The Reykjavik Session
The Ottawa session had decided that the IWGMP should meet again and should attempt to draft a convention on ocean dumping before the Stockholm conference. In the words of the meeting report, the session held in Reykjavik on April 10–15, 1972, was convened "in the hope that agreement on concrete global action might be reached before the [Stockholm conference]" ("Report of the Intergovernmental Meeting on Ocean Dumping," IMOD/4 (April 15, 1972), p. 2). Because regional cooperation appeared unlikely in view of the complexity of the problems and the short time available for negotiation, the meeting was convened under the more indicative working title "Intergovernmental Meeting on Ocean Dumping." It was attended by representatives from 29 states and by observers from the FAO, the IMCO, and the IAEA.[32]

The meeting established a drafting group, which was presented with a negotiation text consisting of the draft convention proposed by the United States, draft articles produced at the previous meeting, the Oslo Convention, and draft articles proposed by Canada. To prevent pollution of the sea, the United States suggested that "the Parties pledge themselves to take all feasible steps." The Oslo Convention and the text from the previous meeting agreed that "the Contracting Parties pledge themselves to take all possible steps." Canada proposed that "Parties pledge themselves to prevent the pollution of the sea" ("Composite Articles on Dumping from Vessels at Sea," Canadian Delegation, April 7, 1972). After considering the various drafts and the proposal of the drafting group, the meeting reached agreement on this formulation of article 1: "Each Party pledges itself to use its best endeavours to prevent the pollution of the sea by matter that is liable to cause harm to the marine environment and its living resources, hazards to human health, hindrance to marine activities including fishing, impairment of quality for use of sea water, or reduction of amenities." (Text of Draft Articles of a Convention for the Prevention of Marine Pollution by Dumping, IMOD/2 (April 15, 1972), p. 3) Article 2 said: "The Parties shall take effective measures individually, according to their capability, and collectively to prevent marine pollution caused by the dumping of harmful matter and shall harmonize their policies in this regard." (ibid.)

The phrase "to use its best endeavors" was obviously less binding than any phrase examined by the drafting group. It was used because representatives of developing countries were opposed to general provisions that they thought would impose unacceptable constraints on their economies. Though developing countries supported the most stringent possible annexes and criteria for dumping permits, they opposed general provisions that proclaimed an overall restrictive and prohibitory dumping policy. Essentially, while developing countries favored stringent standards, they wished to avoid obligations that might hinder their industrialization. Negotiators expected this conflict to resume at Stockholm (memo, Ministry of Foreign Affairs, Denmark (May 1, 1972), p. 7).

The question of which substances would be completely prohibited to dump ("blacklisted" substances) and which substances might be dumped under certain conditions ("graylisted" substances) was from the beginning seen as being of fundamental importance for the convention. Because of the technical nature of such decisions, a working group on the annexes, composed of specialists mainly representing the industrialized countries, was established. The deliberations of the working group were later debated in the general meeting. High-level radioactive waste was allocated to the black list (officially Annex I). This meant a total prohibition on dumping. Medium-level and low-level radioactive waste were not mentioned. The inclusion of high-level radioactive waste and "agents of biological and chemical warfare" in the black list was fiercely contested.

Although the revised U.S. draft convention was silent with respect to dumping of radioactive waste, it was the position of the United States that such activity should not be regulated by the future global ocean dumping convention but instead should remain under the IAEA.[33] In contrast, the draft article proposed by Canada listed high-level radioactive waste among the prohibited substances.

In the subsequent debate, Canada, with strong support from all the developing countries, attempted to include high-level radioactive waste in the black list. The United States and Britain opposed this and argued that the measures to protect against radioactive contamination, as they put it, were best taken through IAEA (memo, Ministry of Foreign Affairs, Denmark, December 6, 1972, p. 3). They proposed that radioactive waste be referred to only in general terms. In their view, an article similar to article 14 of the Oslo Convention (which contained the only mention of radioactive waste in that convention) should be drafted.[34]

In the long debate that ensued, the developing countries were supported by Spain and Portugal; the other Western European states chose not to comment on this controversial matter. The Canadian compromise that was eventually adopted put high-level radioactive waste in brackets (indicating that the issue still was outstanding) and copied article 14 of the Oslo Convention (implying that regulation of radioactive waste should be done through the IAEA). In this way, the draft made a general commitment to take measures against pollution from radioactive waste. However, radioactive waste would not be covered by the operative part of the convention. This diplomatic compromise did not even mention the IAEA.[35] In the end, the Reykjavik session issued a "Text of Draft Articles of A Convention for the Prevention of Marine Pollution" instead of a "Text of Draft Convention for the Prevention of Marine Pollution." Evidently, the session failed to resolve the disagreements (memo, Ministry of Foreign Affairs, Denmark, August 11, 1972, p. 2).

In summary, the question whether to regulate ocean dumping of radioactive waste under the convention was still unsettled. Because the U.S. delegation thought that regulating radioactive waste and bacteriological and chemical weapons in an inappropriate and unreasonable way introduced a disarmament aspect into the dumping convention, this decision was one of the most contentious issues for the coming Stockholm conference (memos, Ministry of Foreign Affairs, Denmark, May 1, 1972, and December 6, 1972). Many reservations on the part of the developing countries, Canada, and (to a lesser extent) Spain and Portugal also promised severe obstacles to agreement at Stockholm. Negotiators expected developing countries to fiercely support stringent standards and at the same time oppose any hindrances to their own industrialization (memo, Ministry of Foreign Affairs, Denmark, May 1, 1972, p. 7). Furthermore, it seemed doubtful that the Stockholm conference, with the participation of a considerably larger number of developing countries and Eastern European states, would be able to reach agreement.

The Second London Session
As the Stockholm conference was rapidly approaching, time became scarce. Once again a session of the IWGMP was called for the purpose of resolving outstanding issues. This session took place in London on May 30 and 31, 1972. The sole working document was the text drawn up at Reykjavik in April. Seventeen states sent delegates; Canada was represented by an observer.[36] No representatives from United Nations or other international organizations were present.

The outstanding disagreements were not resolved. However, an agreement on a proposed new text of the black list that listed high-level radioactive wastes was reached. In addition, the precise definition of this waste would be prepared by IAEA.[37] The Stockholm conference was less than two weeks away, and many states wished to postpone the conditions of the proposed convention until after Stockholm. A number of states stressed the importance of awaiting the conclusions of the forthcoming conference before any further meetings were set. Britain announced a plenipotentiary meeting to sign the convention in late summer of 1972.

The Stockholm Conference
Popular ecology themes from the September 1968 Biosphere Conference, treated in chapter 4, dominated the Stockholm conference's view of nature-society relationships. But the large number of participating states from North and South and the attendance of 550 nongovernmental organizations and many individuals representing mass movements and special-interest groups dramatically distinguished this conference from previous UN-sponsored environment conferences. Furthermore, the Stockholm conference's resolutions and recommendations urged that concrete international machinery for environmental protection be established. Thus, the Stockholm conference was a political conference with "the full rigor of diplomatic protocol" (Hawkes 1972a, p. 736).

The issue of marine pollution control attracted considerable attention at this conference (June 5–12, 1972). According to the British scientific journal *Nature*, this issue was "every delegation's favorite cause," and the chief delegates of Britain and the United States urged action on ocean dumping ("Politics, Bureaucracy and the Environment," *Nature* 237 (June 16, 1972), p. 364). The draft convention and general marine pollution principles were dealt with by Committee III (Pollution and Organizational Matters), but the draft convention on ocean dumping was not a subject of any substantial negotiation. It was evident that several countries wanted more time to study the draft, and that no international agreement would be signed into international law.

But marine pollution was addressed on a more general level. One principle of the Human Environment Declaration from the conference and eight detailed recommendations of the Stockholm Action Plan dealt specifically with marine pollution. The recommendation on ocean dumping urged immediate action: "Refer the draft articles and annexes contained in the report of the inter-governmental meetings at Reykjavik, Iceland, in April 1972 and in London in May 1972 . . . to a conference of Governments to be convened by the Government of the United Kingdom of Great Britain and Northern Ireland in consultation with the Secretary-General of the United Nations before November 1972 for further consideration, with a view to opening the proposed convention for signature at a place to be decided by that Conference, preferably before the end of 1972." (recommendation 86 (d) of the Stockholm Conference Action Plan as approved by the United Nations General Assembly, reprinted on p. 174 of Rowland 1973) Governments clearly felt pressure to demonstrate willingness to act. "For all their differences," one commentator noted in the *New York Times*, "114 countries felt it necessary to show concern for the environment. They agreed on a large number of recommendations, such as an end to whaling and the regulation of ocean dumping, that are useful if not binding. They began the creation of new international machinery." (Lewis 1972)[38] Substantial negotiations would take place at the London Conference. Only a few concrete comments were made on the draft convention when the UN's Seabed Committee met in Geneva shortly after the Stockholm conference. Nonetheless, developing countries repeated that they were vehemently opposed to global pollution control standards.[39]

The London Conference
Representatives from 92 states met in London between October 30 and November 13, 1972.[40] Momentum for an ocean dumping convention was building. The Stockholm conference had generated significant interest, and socialist and developing countries were better represented than at earlier meetings. The British Secretary of State for the Environment, opening "The Inter-governmental Conference on the Convention on Dumping of Wastes at Sea," urged the negotiators to reach agreement on the convention, which would be "the first tangible fruit of Stockholm" ("Conference Meets on Dumping of Waste at Sea," *Nature* 240 (November 3, 1972), p. 4).

The final version of article 1 said: "Contracting Parties shall individually and collectively promote the effective control of all sources of pollution of the marine environment, and pledge themselves especially to take all practicable steps to prevent the pollution of the sea by the dumping of waste and other matter that is liable to create hazards to human health, to harm living resources and marine life, to damage amenities or to interfere with other legitimate uses of the sea." The content of this article should be seen in the light of article 2: "Contracting Parties shall take effective measures individually, according to their scientific, technical and economic capabilities, and collectively, to prevent marine pollution caused by dumping and shall harmonize their policies in this regard."

Thus, beginning in Reykjavik, the general provisions took an even less prohibitory and restrictive stance: states should "promote the effective control" (a less binding formulation) "of all sources of pollution of the marine environment," taking "all practicable steps" to tackle pollution of the sea by dumping. In short, concrete measures should be taken only with respect to ocean dumping. Furthermore, "practicable steps," an elaboration of the phrase "according to their capability" from Reykjavik, implied that regulatory efforts of states would be based on individual technical possibilities and on other factors, especially economic capabilities (memo, Ministry of Foreign Affairs, Denmark, December 6, 1972, p. 5).[41] Similarly, article 2, which also had been redrafted several times during the negotiations, had a new modifier: "according to their scientific, technical and economic capabilities." Developing countries had proposed and strongly supported the phrases "to take all practicable steps" and "according to their scientific, technical and economic capabilities." Instead of setting uniform global pollution standards, their intention was to lessen the burden of dumping reduction on developing countries. As a result, "double standards" were established.[42]

The black and gray lists were evaluated by a group of experts from Canada, France, Indonesia, Kenya, Mexico, Spain, Tunisia, Britain, the United States, and the Soviet Union. They were asked to examine only scientific and technical aspects of ocean dumping, and not address whether radioactive waste and bacteriological and chemical weapons should remain on the black list ("Report of Technical Working Party," DWS(T)7 1st Revise (3 November 1972), p. 1; memo, Ministry of Foreign Affairs, Denmark, December 6, 1972, p. 2). They examined both the draft article from Reykjavik (which had put high-level radioactive waste in brackets) and the alternative formulation from London. They proposed an amended version of the latter, and they agreed that states should take the recommendations of a competent international organization into account when they blacklisted high-level radioactive waste.[43] The IAEA was not explicitly mentioned, although the group agreed on the amendment after it had "fully considered" an IAEA report on the subject (DWS(T)7 1st Revise, p. 3). Significantly, the technical working group now put medium and low-level radioactive wastes (which were not on the black list) on the gray list. Thus, the group proposed that these wastes might be dumped when a special permit was issued and the recommendations of the competent international body in this field were followed. But again, the group did not designate the competent international body.[44]

It was subsequently agreed by the IWGMP that dumping of radioactive waste which the IAEA found unsuitable for dumping for public health reasons, biological reasons, or other reasons was prohibited. Apart from the strengthened role of the IAEA, the final convention text of the black list and the proposed amendment from London were nearly identical.[45] The gray list now included a clause stating that radioactive waste, which was not subject to regulation by Annex I, might be dumped when a special permit was issued and IAEA guidelines and recommendations were followed.[46] In retrospect, the issue of regulation of ocean dumping of radioactive waste had been more of a hindrance to completion of the convention at an earlier phase of the negotiations. The final agreement—clearly a necessary solution to a very contentious issue, and perceived as such—meant that important regulatory decisions remained with the IAEA (memo, Ministry of Foreign Affairs, Denmark, December 6, 1972, p. 19).

Negotiators felt that high international expectations put on their shoulders by the Stockholm conference pressured them to solve outstanding jurisdictional issues from earlier meetings. One U.S. negotiator explained that the fact that "everyone was anxious to complete an effective Convention" resulted in a clause[47] which stated that the convention would not prejudice the Third United Nations Conference on the Law of the Sea (UNCLOS III), which was to begin in 1973 (Leitzell 1973, p. 512). A second U.S. negotiator pointed out that keen attention by the press had had a decisive influence on the final outcome: "Ultimately, the text of an agreement was initialed by representatives of 61 nations, but only after the newspapers had reported, accurately enough, that negotiations were on the verge of bitter collapse, and only after the conference was extended for three days." (McManus 1973, p. 29) The Danish delegation report emphasized the weight of international expectations and newly emerging international norms for environmental protection: "That the negotiations—in spite of all difficulties—were concluded with a signed draft of convention is undoubtedly due to the fact that all participants—also the Soviet Union—felt committed by the recommendations and declarations of the Stockholm conference, whose first concrete result has now manifested itself." (memo, Ministry of Foreign Affairs, Denmark, December 6, 1972, p. 23, translated by L. Ringius)[48]

Conclusions

It conforms well to the neoliberal approach that the United States played an important role when countries decided to cooperate globally on ocean dumping control. The United States took the initiative in this issue area and suggested establishment of a global regime. But, as this chapter also showed, the construction of this environmental regime does not confirm neoliberal propositions about hegemonic power and leadership. The hegemon did not link issues in order to facilitate agreement, nor did it use side payments to attract less enthusiastic states. Rather, as detailed in the previous chapter, U.S. leadership was based on willingness to "take the first steps" to deal with the problem.[49] Significantly, the United States and Britain failed to dictate the regulatory arrangement for radioactive waste, an issue that was dealt with as part of a more comprehensive negotiation package.

Some neoliberal propositions about individual leadership are confirmed by the case. The scientific uncertainties concerning the environmental impact of ocean dumping and significant differences in the willingness and capacity of states to protect the marine environment created significant barriers to agreement in the London Convention negotiations. Thus, to reach agreement, policy entrepreneurs inside the Stockholm secretariat followed a multifacted strategy. In addition to helping governments identify areas of common concern and environmental problems that could be solved through international cooperation, they fashioned and advocated a treaty form that could expedite international environmental negotiations, tackle the problem of insufficient knowledge, separate science and politics, provide a gradual piecemeal approach to ocean dumping regulation, and accommodate the interests of both developed and developing countries—namely, the system of black and gray lists.

International pressure was an important element in reaching agreement on the regime. World public opinion pressured states to act and focused international expectations on an area of possible global agreement. As suggested by the approach concerned with transnational coalitions of policy entrepreneurs, the next chapter will document that the Stockholm secretariat intensively mobilized global public opinion and international pressure during the preparations for the Stockholm conference.

7
Explaining Regime Formation

Building a global environmental regime is a complex and dynamic process. Power, interests, and knowledge might all influence how, why, and under what conditions regimes are built. One combination of power, interests, and knowledge might explain the development of one type of regime dynamics; a different combination might explain the development of a different type of regime dynamics. The three prominent approaches to regime analysis that I presented in chapter 3 analyze regime formation through different conceptual lenses and disagree as to which aspects, steps, and stages are most important.

In this chapter I discuss and compare to what extent the empirical findings presented in chapters 4–6 confirm the hypotheses of power-based, interest-based, and knowledge-based theories as well as those of the approach concerned with transnational coalitions of policy entrepreneurs. In chapter 4 I examined the stages of regime formation that epistemic-community theorists address in most detail. In chapters 5 and 6 I examined the stages that the power-based and interest-based approaches address in most detail. Although I examined different aspects of and different stages in the construction of the global ocean dumping regime in chapters 4–6, and drew some conclusions about the explanatory power of each theoretical approach, in this chapter I systematically compare the four approaches in light of all the empirical findings. This allows me to avoid theoretical straw men and to empirically assess and compare competing hypotheses about regime formation. I conclude in this chapter that a transnational entrepreneur coalition was essential in the construction of the global ocean dumping regime. Only to a small degree did the process of regime establishment unfold according to the propositions of prominent regime theories.

It is useful in a heuristic sense to demonstrate that the prominent approaches to regime analysis, either singly or in some combination, do not account well for the how the global dumping regime was built. But disproving existing theories does not necessarily prove that a different type of regime dynamics exists. Power-based, interest-based, and knowledge-based theories might not account for this particular case, but does that necessarily mean that a qualitatively different type of regime dynamics exists? In this chapter I present more evidence supporting the propositions about the influence of public ideas and transnational coalitions of policy entrepreneurs I developed in chapter 3. Although only a single case study, the analysis is useful in order to detail and carefully explore various kinds of interactions between ideas and policy entrepreneurs and thus to "stimulate the imagination toward discerning important general problems and possible theoretical solutions" (Eckstein 1975, p. 104) in studies of regime formation. I also expect the findings to be relevant to comparable issue areas, and the identified causal relationships to operate in those issue areas.

Knowledge-Based Regime Analysis

The knowledge-based regime approach focuses primarily on how policy changes in response to cognitive and perceptive changes among governmental policy makers. In the environmental field, scholars favoring this approach have focused mostly on the role of scientists and epistemic communities in spreading and communicating new ideas to policy makers. Peter Haas (1992a, p. 27) has argued that epistemic communities are "channels through which new ideas circulate from societies to governments as well as from country to country."

In the case of the global ocean dumping regime, reflectivists would correctly predict that a change in perception of the health of the oceans preceded global regulation and institution building. As the previous chapters showed, a change in perception did take place: the oceans, previously perceived as robust and perhaps even indestructible, were in the early 1970s perceived as being vulnerable, fragile, and endangered. To be sure, global regulation would not have been established had this change in perception not occurred.

It is equally apparent, however, that this new perception of the vulnerability of the oceans was not an accomplishment of a group of ecology-oriented epistemic scientists persuading and pressuring governmental decision makers in a number of counties to act against ocean dumping. The global ocean dumping regime was created not in response to significant improvements in scientific knowledge of the damaging effects of ocean dumping on the marine environment but in response to the powerful idea of "dying oceans" and to new environmental values and beliefs. Scientific uncertainty prevailed in regard to the environmental effects of ocean dumping.

In chapter 4 I noted that members of the U.S. Congress were given conflicting scientific advice in the early 1970s. Congressional hearings revealed deep disagreements between one group of experts, who denied that waste disposal represented a serious danger, and another group (mostly ecologists and marine biologists), who asserted that the limited assimilative capacity of the oceans should be protected by international regulation. Domestic ocean dumping regulation was not, therefore, established by policy makers responding to persuasion and pressures from a unified epistemic community. Moreover, this case confirms neither the notion of epistemic influence and its emphasis on consensual knowledge nor the implicit assumption that within each single policy area or issue area one unified epistemic community advises and influences decision makers. Surprisingly, the epistemic-community literature has often ignored that scientific communities and experts might be offering conflicting advice to policy makers.[1] Therefore, the literature does not throw light on the selection or rejection of policy ideas in situations where scientists disagree.

The politics of expert advice is not well captured by knowledge-based theory either. Two groups of experts competed for policy influence. By powerfully communicating their advice and ideas to the public and to decision makers, environmentalists and "visible scientists" effectively neutralized the policy advice of disagreeing scientists.[2]

Testimony given before a congressional hearing in 1972 illustrates the dominance of the pro-environmental viewpoint. Before explaining his view of the ocean dumping problem to the congressional committee, a marine geologist, who was opposed to a complete ban on ocean dumping, said: "I fully recognize that this approach, as in my statement here, favors ocean disposal of all of certain types of wastes may seem contrary to everything you have heard or read regarding waste disposal at sea." (Smith, "Statement," in U.S. Senate, Ocean Waste Disposal, p. 206) This illustrates well that a community of professional waste managers and its supporters had been unable to successfully challenge claims made by more publicly visible and vocal scientists and ecologists.

The anti-dumping view dominated and was advocated by policy entrepreneurs in government and backed by ecologists and environmentalists influencing the public and the government though the mass media. As will be elaborated below, the idea that "the ocean is dying" came to be seen as being morally superior to all other views, and this further increased its influence and reduced the influence of disagreeing scientists. But Peter Haas and his colleagues tend to overlook that epistemic communities need to make convincing arguments that appeal to decision makers as well as to the public in order to influence policy. It is necessary to compare epistemic communities' influence with the influence of policy entrepreneurs and other ideas-based actors, including the role of the media. Moreover, these environmental regimes analysts mostly ignore national and transnational policy entrepreneurs and claim that it is predominantly scientists who place environmental issues on the public agenda.[3] Yet scientific and epistemic communities are clearly not the only channels through which new beliefs, ideas, and knowledge about the state of the environment circulate from societies to governments and among countries. A more satisfactory knowledge-based approach would have to identify with more precision the nature of processes determining the selection of ideas and the conditions under which epistemic communities affect policy.[4]

Peter Haas concludes in his analysis of the Med Plan that a shared, or at least nonrival, perception of the environmental problem among developing and developed countries facilitated the establishment of a regime. However, as was demonstrated in chapter 6, the global ocean dumping regime was successfully built despite the fact that relatively few developed countries saw ocean dumping as a significant international problem, while developing countries generally thought that developed countries had created the problem and that they therefore should solve it. As will be further documented below, policy entrepreneurs inside the Stockholm secretariat, not a transnational network of scientists, helped governments overcome serious obstacles to regime establishment in the absence of a shared perception of the environmental problem with which they were dealing.

The swiftness of this regime-formation process should be noted.[5] It could explain the relative insignificance of scientists and experts as well as the demand for leadership by skillful policy entrepreneurs in this case.[6] Less than 2 years elapsed between when the United States first defined ocean dumping as an area that would benefit from cooperation and when the global ocean dumping regime became a reality. Presumably, no priority was given to development of a scientific basis for ocean dumping policy because it was widely perceived that this problem was relatively urgent and significant. Developing a scientific basis for decision making would have delayed the policy-making process and was neither necessary nor sufficient in order to push the regime-formation process forward. In the Med Plan case, in contrast, it took much longer to build the regime and, because more weight was put on marshaling scientific evidence, networks of scientists were better positioned to play a significant policy role.

Ernst Haas (1990, p. 226) has suggested that the leadership style of Maurice Strong was based on using a "privileged body" of scientific and technical knowledge to persuade governments. Strong and his staff did indeed perform crucial policy entrepreneurship in the formation of the global ocean dumping regime and in the time of the Stockholm conference. Significantly, the secretariat was well informed about states' interests and positions with respect to ocean dumping, and it could therefore design and propose regime features that helped states overcome significant obstacles to cooperation. As is true of the domestic policy process in the United States, persuasion through use of exceptional scientific documentation of environmental damage was not an essential facet of the entrepreneurship of Strong and the Stockholm secretariat.

In summary, knowledge-based hypotheses about formation of environmental regimes do not explain this case well. Despite presenting a case study that is comparable to the Med Plan and other epistemic community-oriented case studies, this study of the influence of marine scientists on the regime-building process does not confirm key knowledge-based propositions.

Power-Based Regime Analysis

Realists would assume that the United States, the hegemon, would build the global ocean dumping regime to protect its national interests. They would stress, quite correctly, that the global ocean dumping regime was an expansion of U.S. domestic regulation onto the global level. This group of scholars would further point out that, since the regime potentially would harmonize costs of pollution control across countries, and thus would neutralize economic costs imposed on U.S. industry by unilateral domestic regulation, a strong economic incentive existed to act as a leader in the regime-building process. In short, realists could claim that the regime was created by a hegemon acting in its own interest. Apparently, this case would provide no new insights into regimes, since "the prevailing explanation for the existence of international regimes is egoistic self-interest" (Krasner 1983, p. 11). Self-interest could also explain why developing countries were opposed to international regulation entailing economic costs for their economies, and why large nuclear nations would try to protect their regulatory autonomy with regard to radwaste disposal by resisting the regulation of this dumping under the regime.

Governments did not, however, reach agreement out of concern for their own interests as traditionally understood by realists. The regime did in fact not reflect an overwhelming concern for protecting national welfare. In the words of the economist Charles Pearson (1975b, p. 80), "it appears that the architects of the agreement had more than a narrow national welfare perspective in mind and, indeed, had some attachment to a global welfare concept." Brazil refused to sign the dumping convention in London because it did not go far enough in reducing ocean dumping. The head of the Brazilian delegation told the press that the convention "defends the interests of the developed countries, polluters every one" (Hawkes 1972c). Sweden followed a policy of strict control of dumping at home as well as abroad, and President Nixon had signed U.S. ocean dumping legislation to protect the Atlantic before the London Conference. Britain had already joined a regional agreement protecting its waters against ocean dumping (the so-called Oslo Convention, covering a large part of the North Sea area), so Britain's decision to join the global ocean dumping regime cannot be understood as motivated only by concern for its own waters. As the evidently proud British Under-Secretary of State for the Environment declared at a press conference after the agreement was reached in London in November of 1972: "It is an important step in controlling indiscriminate dumping in the sea. Nations are now going to take effective control not only of their territorial waters, but of the whole sea." (Bedlow 1972) The international press quoted Russell Train, leader of the U.S. delegation in London, as declaring that the convention represented "a historic step toward the control of global pollution" ("Saving the Seas," *New York*

Times, November 22, 1972; "Policing the Dumpers," *Newsweek*, November 27, 1972). Train further said that the conference had met the goal set by the Stockholm conference, and that the agreement provided "practical evidence of the increasing priority the nations of the world are giving to environmental problems" (Arbose 1972).[7]

In chapters 4 and 5 I noted that realists would have to include the domestic level in order to understand how the initiative to establish the global ocean dumping regime grew out of U.S. domestic regulation. But realist theory has not developed a theory of domestic politics that can be examined in this case study. Moreover, little is said in realism about the roles of ideas, policy entrepreneurs, and nonstate actors, or about international organizations. A further complicating factor is that, while preserving territorial and political integrity are core objectives of a hegemonic state, and, for realists, serve the national interest, protection of the oceans against dumping can hardly be seen as closely associated with such objectives. This complicates the realist analysis, although scholars sympathetic to realism have suggested using an inductive approach to analyze policies that are not closely associated with preserving territorial and political integrity as examples of protection of the national interest.[8] But, as I noted in chapters 4 and 5, realists following an inductive approach would still need to acknowledge that the hegemon perceived ocean dumping as an essentially global problem, and that dealing effectively with ocean dumping demanded cooperation among states. Similar to the knowledge-based approach, the power-based approach cannot account for why the United States and a number of other Western countries saw ocean dumping as an international problem of some significance and urgency in the early 1970s. More generally, this case cannot be explained by pointing to states' interests; the way states defined their interests was changing. To fully comprehend this process of regime formation, it should be taken into account that the view of the health of the oceans changed significantly in the beginning of the 1970s, especially in the West.

Evidence of the changing view of the oceans abounds. A British scientist who was at the time the editor of the highly respected journal *Nature* wrote in 1972 that incidents such as the recent discovery of large amounts of mercury in Pacific tuna had helped to create "the impression that the oceans of the world are in some general sense polluted and that the pollution stems from industrial activity of a kind which is characteristic of the twentieth century" (Maddox 1972, p. 117). The prominent economist and ecologist Barbara Ward vividly described the same change in ideas in one of a series of lectures[9] held in Stockholm during the UN conference: "One of the fascinating things about the present moment is the speed with which truth is moving toward platitude. There are ideas and concepts which, when I wrote them in our preliminary draft last year, made me wonder how far out I could be. Yet today Ministers of the Crown are saying them and that is surely about as far in as you can get. . . . In today's debate . . . delegates talked above all of the vulnerability of the oceans. Yet only a year ago, this was an entirely new idea. Now it is a *lieu commun*, a near-platitude. . . . The new ideas are penetrating human consciousness with incredible rapidity." (Ward 1973, pp. 21–22)

Ocean dumping in particular was seen as essentially an indivisible international public "bad" because ocean currents were thought to mix wastes and transport them great distances. As an insider in the U.S. marine scientific community noted, a new international view on this issue emerged about 1970: "There unfolded an awareness that waste of national origin dumped at sea may be distributed globally. While such threats were not regarded as immediate or of crisis proportions, the pervasiveness of the fluid media potentially exposed all nations to the same risk and uncertainty. So whatever the geopolitical and geoeconomic considerations in debate, no matter how parochial the arguments, participants came to recognize that all questions shared a central core of scientific, technical, and economic facts not constrained by political or institutional boundaries or ideology." (Wenk 1972, p. 425) In another example, one senator said immediately before the Senate passed the ocean dumping bill: "The oceans have currents, just like the rivers, as we know, so the debris and waste going into the oceans from Western European countries, Japan, and any industrialized nations, finds its way to the shores of this land, just as the debris and waste which we put in the oceans along our coast finds its way to London, Stockholm, and other parts of the world." (*Congressional Record: Senate*, November 24, 1971, p. 43071) This idea bound states together; no state could solve the problem alone. Realism's self-interest, as opposed to common or shared interest, cannot satisfactorily explain this process of regime formation. In short, power-based explanations are unable to account for the United States' environmental interests, concerns, and priorities with respect to the ocean dumping regime.

Power-based explanations would also have difficulties accounting for how the global ocean dumping regime was built and for how and why states managed to reach agreement. Realists claim that regime formation would be very unlikely in a situation of significantly opposed interests among developed and developing countries. As was discussed in chapter 3, they would predict that the hegemon, perhaps together with other powerful states, would have to use force, side payments, or manipulation of incentives to build the regime.

In this regime-formation process, however, U.S. leadership was based on ideational power. The United States attempted to initiate cooperation by demonstrating to other nations the advantages of cooperation. Political and administrative leaders emphasized that the ocean dumping problem had to be dealt with effectively at home before international negotiations were initiated. As a representative of the Council on Environmental Quality said before a congressional hearing: "If the United States is in fact to exercise leadership in this critical area, if it is to persuade other nations to control their ocean disposal of wastes, then it is essential that the United States first put its own house in order. In my opinion, prompt and favorable action by Congress to establish effective regulation of ocean dumping is a prerequisite to action by other nations." (Gordon J. F. MacDonald, "Statement," in U.S. Senate, Ocean Waste Disposal, p. 49) Similarly, a senator declared just a few weeks before government representatives met in the London Conference: "If the United States does not take the affirmative action to secure authority to control ocean dumping from its own shores, we cannot expect to be persuasive or to maintain the credibility of our leadership on this issue at the international meeting in London." (Sen. James Buckley, *Congressional Record: Senate*, October 3, 1972, p. 33311) Power-based theories, however, are concerned with leadership through coercion and force.[10] No attention is paid to leadership based on persuasion and readiness to deal with a common threat or a collective problem.

It seems quite clear, finally, that U.S. leadership was a necessary but not a sufficient condition for the construction of the global ocean dumping regime. Chapter 6 showed that policy entrepreneurs active inside the Stockholm secretariat played a crucial role in the negotiations. But realists' preference for system-level explanations diverts attention away from policy entrepreneurship by high officials of international organizations.

Interest-Based Regime Analysis

It conforms well to the interest-based approach that the United States took the lead when the global dumping regime was built. However, the hegemon did not make use of its superior material resources, nor did any individual representing the hegemon act as an outstanding entrepreneurial leader. Rather, as was pointed out above, the United States' leadership strategy relied on persuasion. The United States wanted to set a good example for others to follow—a form of leadership that is distinguishable from structural, entrepreneurial, and intellectual leadership.

Unilateral action demonstrating determination and showing that national policy priority is given to solving a collective problem clearly has a demonstration value and is rightly understood to be a means of persuasion. Yet many neoliberals would doubt that a state might provide leadership in regime formation by going first and by behaving in an exemplary fashion. One exception is Arild Underdal, who observes (1994, p. 185) that "unilateral action may provide leadership . . . through social mechanisms, notably as a means of persuasion. Particularly in situations characterized by high problem similarity, unilateral action may be used for demonstrating that a certain cure is indeed feasible or does work, or to set a good example for others to follow." Underdal concludes that "the persuasive impact of unilateral actions depends primarily not on its actual impact but rather on the amount of uncertainty removed [about an actor's motives and strategies] or on its moral force and symbolic significance" (ibid., p. 185).[11] These observations are most pertinent in the present case, but persuasion understood as a social mechanism and the moral and symbolic power of state behavior are not significant research issues in interest-based regime analysis, nor are they well integrated in its research program. For example, Keohane (1984, p. 53) claims that strategies relying on persuasion and the force of good examples are less effective in achieving cooperation than strategies involving threats and punishments as well as promises and rewards.[12] The role of processes of persuasion and communicative action are, however, fundamental issues in the ideational approach to political analysis.

Rationalists generally expect large groups to reach agreement only with great difficulty. Collective action theorists posit that a large group of governments encourages governments to free ride because the group will provide the collective good despite free riders who therefore are not strongly motivated to join the group. A different pattern of behavior, however, is observed in this case: a large group of governments increased the pressure on individual governments to cooperate in a situation where international public opinion encouraged collective action. Discussing how voluntary movements succeed to provide public goods when Olson's collective action theory would predict their failure, it has similarly been noted that "sometimes the mass media help to build up the equivalent of small-group, face-to-face social pressures" (Douglas and Wildavsky 1982, p. 116).

In regard to regime formation more broadly, several of Oran Young's propositions are confirmed in the case of this regime. First, integrative bargaining dominated over distributive bargaining. Second, despite objections by some countries (most prominently Brazil), it seems quite clear that a large measure of justice was achieved in the end. Third, the approach taken in the convention was rather straightforward and productive, supporting the importance of salient solutions. Fourth, the case evidently confirms the claim that it is necessary that states be guided by concern for joint interests and common goods.

The case offers less support for the proposition about importance of shocks or crisis as it is necessary to differentiate between developed countries (which generally were concerned about the ocean dumping problem) and developing countries (which generally were much less concerned or even unconcerned). Somewhat similarly, the first of the two propositions about policy priority (that high policy priority is necessary for achieving agreement) is confirmed in the case of developed countries, but the second proposition (that, conversely, low priority makes it easier to reach agreement) is supported in the case of the developing countries. The importance of establishing credible compliance mechanisms is neither confirmed nor disproved. And, unless interpreted quite broadly, the three probabilistic propositions hypothesizing that focusing closely on the scientific and technical basis of the regime facilitates negotiation are not supported in this case.

As was described in chapter 4, one international conference sponsored by the U.S. Congress dealt specifically with ocean dumping, and several other conferences and international meetings brought together foreign decision makers, prominent environmentalists and ecologists, scientists, officials of international organizations, and the Stockholm secretariat as part of the United States' preparations for the Stockholm conference. This fits well with the interest-based approach, as it pays considerable attention to international organizations and transnational coalitions. It also introduces more actors into the process through which states' interests are defined than do knowledge-based and power-based approaches.

Some neoliberals have paid much attention to individual representatives of international organizations acting as leaders when regimes are built. This is indeed pertinent, since the Stockholm secretariat was deeply involved at several important stages of the regime-building process and played a crucial role in moving governments toward agreement. As was detailed in chapter 6, policy entrepreneurs within the secretariat developed and promoted solutions that removed significant obstacles to a successful completion of the negotiations. Thus, this case evidently supports the claim that entrepreneurial activities are necessary for a regime to form. But policy entrepreneurs also provided a substantial measure of ideational leadership and mobilized international public and political support over the entire process of regime establishment; indeed, ideational power was not confined to the problem-definition and pre-negotiation stages. Taken together, these activities—raising issues, brokering among stakeholders, and mobilizing and broadening support—seem better understood as manifestations of policy entrepreneurship as defined in chapter 3. Young's propositions about structural, entrepreneurial, and intellectual leadership neglect other, equally important facets of policy entrepreneurship that were influential in this case. Whereas the case disproves Young's strong hypothesis that the presence and interplay of at least two of the three forms of leadership is necessary (although not sufficient) for a regime to arise, it supports Underdal's propositions about a possible "optimal mix" of leadership modes in multilateral negotiations.[13]

It has recently been argued that ideas and focal points around which the behavior of actors converges under certain circumstances facilitate cooperation among self-interested actors. Crucially, as Garrett and Weingast emphasize (1993, p. 176), focal points are not automatically given; they must be constructed.[14] But rational choice theory cannot predict why a certain idea is selected over others. This is instead a question for an ideational approach to political analysis. Environmental problems and international issue areas are human artifacts, so perceptions often vary over time and across actors. Only after sustained efforts of persuading, selling, and campaigning by a transnational coalition of policy entrepreneurs were countries' expectations converging in the ocean dumping issue area. The coalition constructed an issue area where none had existed.

As the next section further documents, the Stockholm secretariat was an important policy entrepreneur and had considerable ideational influence on regime formation. The Stockholm secretariat was at the same time a mobilizer, a popularizer, and a legitimizer. But because interest-based theory focuses narrowly on bargaining, negotiators, and directly involved participants, it largely ignores leadership premised upon widely shared values, beliefs, and public perceptions. It overlooks the fact that transnational policy entrepreneurs under certain circumstances are able to shape and mobilize public opinion and to influence governments' beliefs, values, and interests, and the fact that negotiators respond to such changes.[15]

Similar to regime approaches emphasizing power and knowledge, the interest-based approach cannot explain why states redefined their interests with respect to the oceans in the early 1970s, or that a new perception of the impaired health of the oceans emerged at that time. And because this approach is concerned mostly with interstate interactions, it is less helpful in understanding the interactions between domestic and international factors in regime formation, most importantly the U.S. initiative to establish the regime and the U.S. leadership strategy emphasizing communication and international persuasion.[16] It is insufficient, finally, to point to protection of self-interest as the primary or perhaps the only motivation behind the creation of this global environmental regime.[17]

Policy Entrepreneurs and Public Ideas: The Essential Ingredients

First of all, a group of U.S. legislators were determined to "clean up the oceans" despite conflicting expert advice. As was described in chapters 4 and 5, they were not pressured and persuaded by an epistemic community. These policy entrepreneurs allied themselves with prominent scientists and leaders of the environmental movement in order to mobilize and frame public and political opinion and to establish regulation of ocean dumping. They used congressional hearings, the media, and environmental leaders to communicate new ideas. They did not follow public opinion; they led it.

I chapter 5 I quoted from Senator Ernest Hollings's closing remarks at the International Conference on Ocean Pollution: ". . . the only way . . . we are ever going to get this message through is with people with the brilliance and dynamism of you three [Commoner, Heyerdahl, and Hugh Downs] here this morning getting the attention of the American public and in turn of our colleagues here in the Congress to move in the right direction" (Sen. Ernest Hollings, International Conference on Ocean Pollution, p. 26). Members of Congress were keenly aware that Cousteau, Heyerdahl, and other articulate science celebrities and prominent, vocal scientists and ecologists played a critical role in identifying the need for regulation and establishing new norms for ocean protection. In short, they performed a crucial norm-setting role.

Ecologists and environmentalists originally produced the ideas and concepts that bound together the transnational coalition of policy entrepreneurs. But politicians were not passive consumers of ideas. Politicians themselves were actively involved in spreading policy-relevant ideas, and it is difficult to distinguish in a meaningful way among producers, consumers, and communicators of ideas. For example, Senator Gaylord Nelson, who first proposed U.S. ocean dumping regulation in Congress in February of 1970, wrote an article titled "Stop Killing Our Oceans," which was published in *Reader's Digest* (February 1971) and reprinted in the *Congressional Record* (*Senate*, February 8, 1971, pp. 2035–2036).[18]

The case of ocean dumping confirms the claims made in chapter 3 about the significance of the intellectual and contextual characteristics of ideas. Policy entrepreneurs chose to push an idea and target a problem that corresponded with broad environmental concerns, whereas ideas that questioned the seriousness of the marine pollution problem were relegated to political obscurity and survived only among scientists and in their professional journals.[19] The "dying oceans" idea was a strong value statement favoring a general need for protecting the marine environment. It expressed, in capsular form, a more comprehensive environmental value and belief system.

It was a combination of intellectual and contextual characteristics that made the "dying oceans" idea pervasive and powerful.

First, the idea was (or at least could be) used to portray a large number of very different instances and degrees of contamination and pollution. Rivers, lakes, estuaries, straits, seas, and oceans were repeatedly described as "dead" in this period.[20] In addition to the examples mentioned earlier, a part of New York Harbor was called "the Dead Sea,"[21] the Santa Barbara Channel was described as "a sea gone dead" after an offshore oil well blowout in 1969,[22] and the Baltic Sea was said to be on the verge of a disaster.[23] The idea's broad applicability made it highly influential.

Second, the image of the ocean is a powerful symbol of our common heritage (as presumably all species came from the oceans), tranquility, and global interdependence, and thus it is very much antithetical to a world of nation-states jealously guarding their sovereignty.[24]

Third, the oceans' traditional legal status as common property, with its connotations of international, global interests as opposed to the chauvinistic, national interests of states, probably added to the power of this idea.

Fourth, the "dying oceans" idea did not exist in an ideological vacuum. It echoed two of the major themes of the environmental movement in the late 1960s and the early 1970s. One major theme of the environmental movement was the feeling of crisis (Weart 1988, p. 324). There were, in a sense, two crises, because the crisis of the ocean would imperil the welfare of individuals and the health of society. A second major theme of the environmental movement was "a concern with whole systems, with the ways in which pollution could affect ecological relationships around the globe, and with the ways in which our entire modern culture and economic system encouraged pollution" (ibid., p. 325). On a first impression today, ocean pollution would probably appear to be a problem with obvious international and even global dimensions. However, not until the late 1960s did ocean pollution come to be looked upon as an international environmental problem.

By advocating and spreading the "dying oceans" idea, policy entrepreneurs increased the legitimacy and the credibility of environmentalists and pro-environment mass media attacking the prevalent idea of the pristine and indestructible ocean. Policy entrepreneurs staged national hearings and international conferences, and environmentalists' vivid descriptions of how pollution was not only endangering life in the oceans but the entire global ecosystem got prominent coverage in the media. Experts, scientists, and professionals who took a moderate pro-dumping stance were not given equal access to important and prestigious political forums and platforms.

It also conforms well to prior research that policy entrepreneurs deliberately pushed a simplistic understanding of the problem of ocean dumping in order to mobilize political and public support. Evidently, the idea that "the oceans are dying" was not true in an absolute sense, but this did not prevent it from having a tremendous impact on policy development. As little was scientifically known about the environmental effects of ocean dumping, policy entrepreneurs could claim that the oceans were "dying" without being disastrously disproved by others. The mass media, moreover, were very important in communicating policy-relevant ideas and policy proposals concerning ocean dumping to politicians and the public. This is also confirmed by studies showing how politicians use the media to communicate ideas and to focus attention on particular issues.[26] Also, policy makers may often chose to use the media, instead of technical reports and briefing papers, to focus the attention of other policy makers on policy problems.[27]

The "dying oceans" idea held a moral power that was accentuated at the point when public opinion and public debate about policy generally was shifting in favor of environmental protection. The moral power of the idea further added to its impact on problem definition and policy formulation. This was evident in the following remark, made by a marine geologist supporting continuation of controlled ocean dumping at a congressional hearing: "I recognize also that in the present era of aroused public interest in the environment, in which ecology has become virtually a 'motherhood issue,' there are certain significant hazards, both politically and professionally, in what at first may seem to favor what others might term pollution." (Smith, "Statement," in U.S. Senate, Ocean Waste Disposal, p. 206) "The dying oceans," an image that resonated well with recent spectacular examples of dumping accidents, burning rivers, and polluted lakes, thus was a normative statement that endorsed a need generally to protect and preserve the marine environment.

The above illustrates that the "dying oceans" idea, after it had been accepted, exerted a significant influence on the viability and credibility of other ideas. As was pointed out in chapter 3, there is a clear tendency for ideas, once they catch on, to dramatically reduce the attractiveness and sometimes even the legitimacy of alternative ideas. This idea framed the ocean dumping issue in such a way that the environmental and ecological viewpoint seemed credible and morally superior to other views, whereas the view that the ocean could under certain conditions safely accumulate some wastes seemed without credibility as well as "non-environmental." Scientists and experts advocating such a view therefore had negligible influence on the framing of the ocean dumping issue. This is confirmed by sociologists' studies: "At any specific moment," Gusfield (1981, p. 8), noted, "all possible parties to the issue do not have equal abilities to influence the public; they do not possess the same degree or kind of authority to be legitimate sources of definition of the reality of the problem, or to assume legitimate power to regulate, control, and innovate solutions." As policy entrepreneurs politicized the issue, it became increasingly difficult for alternative views to enter and play a role in the policy-making process.

The case of ocean dumping also supports the claim that political feasibility is a major concern when policy entrepreneurs choose which problems to target as well as how they target and frame them. Although land-based sources of marine pollution were far more important from an environmental point of view, policy entrepreneurs realized that ocean dumping was a "tempting target for politicians" (McManus 1983, p. 121): the source was discrete and easily regulated, and no important special-interest groups would mobilize against regulation. As was discussed in chapter 6, policy entrepreneurs inside the Stockholm secretariat found that the higher scientific uncertainty and the higher abatement costs made land-based marine pollution a much less attractive target than ocean dumping.

But more than anything else, the United States' intention to regulate ocean dumping made this issue an obvious target for Maurice Strong and his secretariat. As early as November of 1970, they singled out the issue as a candidate for international agreement and consistently pushed it at the global level. Together with prominent international environmentalists and U.S. legislators, they formed a transnational coalition of policy entrepreneurs spreading the "dying oceans" idea and advocating global control of ocean dumping. The U.S. draft convention established the necessary starting point for negotiations, and, when entrepreneurs inside the Stockholm secretariat established the IWGMP, as one U.S. negotiator put it, "the stage was set for the United States to table a draft ocean dumping treaty" (McManus 1973, p. 29). The secretariat succeeded in focusing negotiations on an area of possible agreement and establishing a useful negotiation forum for the hegemon.

The United Nations was influential in legitimizing international ocean dumping norms. The Stockholm conference heightened international expectations, and the governments meeting in the London Conference felt committed to recommendations agreed to at the conference. This was so because the Stockholm secretariat managed the ocean dumping issue in such a way that the international community's willingness to protect the environment came to be at stake in the negotiations. "And naturally," one experienced negotiator said caustically, "everyone wants to see at least one treaty result from a Conference as heralded in advance as Stockholm is" (Mendelsohn 1972, p. 391).[28] As was pointed out earlier, the convention would establish the first global regime for environmental management to embody the success of the Stockholm conference.[29]

An internal secretariat memo from Peter Thacher to Strong, dated December 20, 1971, detailed the Stockholm secretariat's intense involvement in enhancing the conference's visibility. Moreover, heavy emphasis was put on advocating policies that were flexible and dynamic. Among other things, the memo said the following:

- The primary task between now and June is to refine and sell the product; get negotiations started leading to favourable action by Governments at Stockholm. The Prepcom [Preparatory Committee] session in March should be used to start a planned promotion campaign.
- In order to plan the campaign that is it to get underway at the Prepcom we need to identify major target governments, i.e., those from whom support is critical, and draw up a plan for each government identifying the points of influence, both official and private (including mass media) which should be approached, as well as all assets, "friends," available to assist.
- The plan should also take into account various international meetings at which the promotional material can usefully be presented. With regard to specific subjects such as marine pollution, there may be regional and other meetings of importance at which our participation will be useful. We need to consider whether further work along the lines of our proposals can be advanced before Stockholm.
- Immediately after a final review of experience gained at the Prepcom the planned promotional activity by staff, consultants, and "friends" should be launched.
- Essential to the above promotion campaign is the dynamic quality of the actions proposed; they are not being presented on a take-it-or-leave-it basis, they are subject to modification both before and, if necessary, at Stockholm.

To "sell the product" (that is, the conference's conceptual framework and the proposed actions), the secretariat planned a global "promotion campaign" and prepared and distributed "promotional material." International mass media played a crucial role in focusing global attention on the conference and thereby on the environment, thus increasing domestic pressure on governments.

At Stockholm, prominent international scientists and ecologists invited by the Stockholm secretariat again attacked the view of the oceans as pristine and indestructible. Heyerdahl, who had earlier appeared at congressional hearings and at meetings of the IWGMP, spoke in the distinguished lecture series held concurrent with the Stockholm conference. Using more dramatic statements than he had on previous occasions, he said: "Quiet recently it has become more and more apparent that some of the changes man is imposing on his original environment could be harmful to himself; in fact they could even lead to global disaster.... Since life on land is so utterly dependent on life in the sea, we can safely deduce that a dead sea means a dead planet." (Heyerdahl 1973, pp. 46, 49)

Visible scientists were not just suppliers of scientific data; they also identified particular needs for and legitimized international environmental regulation. Maurice Strong had asked René Dubos and Barbara Ward to prepare the "conceptual framework" for the Stockholm conference. The study, published under the title *Only One Earth: The Care and Maintenance of a Small Planet* (Ward and Dubos 1972), was on sale at the time of the conference.[30] Strong had suggested that the study identify areas in which there "exists a sufficient degree of consensus in the scientific community to permit us to act" ("Statement," in U.S. Senate, International Environmental Science, p. 47). The timing could hardly have been better. This book was telling evidence of the environmental concern of the late 1960s and the early 1970s.

Strong and his secretariat acted as a forceful policy entrepreneur on behalf of the global environment. Mostly Strong, but also other members of the secretariat, met with representatives of the international scientific community, governments, the business community, and environmentalists in various combinations and forums around the world.[31] These meetings created opportunities to mobilize support by spreading environmental ideas, distributing promotion material, and meeting with "friends." Moreover, at these meetings the secretariat learned about the concerns and interests of governments and various stakeholders. This knowledge about the interests of governments and other stakeholders was important and most likely necessary in order to be able to provide influential policy entrepreneurship. In the early 1970s governments were not familiar with environmental issues, had little knowledge of the views, priorities, and interests of other governments, and had little experience with comparable interstate interactions and instances of cooperation. The Stockholm secretariat was more familiar with the concerns and interests of governments and other stakeholders than governments themselves were. A wide policy window appeared, and the secretariat, which was well placed to take advantage of the opportunity, brokered effectively. In summary, the secretariat simuntaneously pushed and pulled governments to cooperate by mobilizing public opinion and support, identifying joint gains to states, devising politically attractive regime features, and forming coalitions with domestic groups.

Conclusions

A transnational coalition of policy entrepreneurs played a major role in the establishment of the global ocean dumping regime. It persuaded states to change or redefine their interests with respect to ocean protection, identified shared interests among states, discovered joint gains, mobilized domestic support, and designed politically attractive features of the regime. In short, the transnational coalition of policy entrepreneurs provided both ideational and entrepreneurial leadership in the regime-building process. The interests and positions of states in the environmental field were recent and developing rather than well established and fixed, and influential policy entrepreneurs knew states' interests and positions better than states themselves did. This informational advantage made it possible to design and propose regime features that helped states overcome significant obstacles to establishment of the global ocean dumping regime. Entrepreneurs could furthermore mobilize domestic support for international cooperation in a situation where new widespread environmental ideas, values and interests were emerging, thus creating common ground among states. They effectively persuaded governments to adjust domestic and international policies in line with a recent fundamental ideational change.

This case documents an ideational path to regime establishment that previously has attracted little attention. As I have shown in this chapter, none of the prominent approaches gives a satisfactory account of how the global regime for controlling ocean dumping was built. To judge from this case, knowledge-based analysis attaches too much importance to the knowledge input of scientists. This case shows no evidence that networks of scientists exerted a strong influence on the framing of the issue of ocean dumping or on the formation of a regime. This case shows that the hegemon, using persuasion rather than force and coercion, provided leadership. The interest-based analysis accounts well for the entrepreneurial leadership provided by Maurice Strong and his secretariat, but it overlooks the influence of ideational factors in environmental policy initiation and regime formation.

Chapter 3 showed that power-based, interest-based, and knowledge-based regime analyses disagree on a number of theoretically significant issues. In addition to those differences, the dominant regime theories have some significant similarities and shared weaknesses.

First, there is widespread agreement that public opinion—hardly mentioned in the literature on regimes—has little or no effect on cooperation for the protection of global commons such as the oceans.

Second, all three approaches largely exclude the domestic level from their analysis of regime formation. As my study has repeatedly demonstrated, examination at domestic levels is necessary in order to understand the origins of the global ocean dumping regime. It is also evident that scientists are just one among many groups potentially influencing domestic policy development.

Third, regime analysis views international organizations, especially when understood as policy entrepreneurs in their own right, as rather insignificant. The epistemic-community approach tends to erase the boundaries between national and international bureaucracies, although it is open to the possibility that an international secretariat might use scientific knowledge to persuade countries and to facilitate negotiation among governments, and furthermore pays attention to bureaucratic power of epistemic communities. The interest-based theory is primarily concerned with how individuals associated with an international organization might use their negotiating skills productively to build regimes. Such approaches ignore other, equally important facets of policy entrepreneurship and the weight of public opinion and international moral pressure.

The United States' leadership in the formation of the regime was ideational in nature. It was not based on coercive measures or on manipulation of incentives. A series of U.S.-sponsored international conferences at which prominent international environmentalists and scientists vividly described the need to act internationally against ocean dumping were intended to influence international public opinion and to change the interests and values of states. The United States, which was first and most concerned about the global marine environmental aspects of ocean dumping, projected its view onto the international arena. Leadership primarily hinged on vivid descriptions of the dangers of this problem and demonstration of willingness to do something about it. The American perception of the ocean dumping problem, which can be traced back to a transnational coalition of policy entrepreneurs, informed the negotiations on the global ocean dumping regime. Few developed countries had individually taken steps to regulate ocean dumping, and most other countries (developing countries included) were far less if at all concerned about controlling marine pollution when the global ocean dumping regime was created. The hegemon, the United States, strongly influenced international norm formation in the ocean dumping issue area.

Although the three prominent regime approaches pay attention to a number of aspects, steps, and stages of regime formation, they ignore the importance of widely shared ideas, values, and policy entrepreneurship. This study shows that detailed examination must include all levels at which regime building takes place. Prominent regime approaches artificially separate domestic and international levels; equally artificially, interests and power are seen as antithetical to ideas and knowledge. A better approach would fit together, not separate, all aspects and actors that contribute to the establishment of a regime.

8
Changing the Global Ocean Dumping Regime

As was noted in chapters 1 and 2, the global ocean dumping regime declared a ban on the dumping of low-level radioactive waste in the world's oceans in 1993. This permanent ban marks at the same time the most significant policy development in the history of this global environmental regime and in the history of international regulation of radwaste disposal. The ban and the forces supporting it terminated radwaste disposal at sea, a practice that had been in use since 1946. Importantly, ocean dumping was suspended although several major nuclear nations—principally the United States, Britain, France, and Japan—had a considerable stake in disposal at sea because they lacked sufficient permanent land-based storage facilities for their low-level radioactive waste.[1]

Radwaste disposal is a significant environmental, security, and energy independence issue for countries that use it. Realists would expect dumper nations to fiercely protect radwaste disposal against attempts at interference by other nations. From a realist viewpoint, it is therefore highly unlikely that a powerful group of pro-dumping nations would accept a termination of radwaste disposal. From a neoliberal perspective, a termination of radwaste disposal is unlikely to happen as long as this waste disposal practice creates benefits exceeding its costs to the dumpers—in other words, as long as dumping low-level radioactive waste at sea makes the dumper better off than discontinuing ocean dumping or other alternatives. In view of the economic and political power of pro-dumping countries, it is most unlikely that anti-dumpers would be able to offer pro-dumping countries an attractive alternative to ocean disposal. Epistemic-community theorists would expect that a transnational network of scientists had persuaded pro-dumping governments to halt radwaste disposal. Governments, uncertain about the environmental and human health effects of radwaste disposal, would follow the advice offered by acknowledged technical experts and scientists. Contradicting these propositions, as I shall demonstrate in this chapter, the actions of a global ENGO were decisive in the regime change, independent of simple state interests, epistemic communities, or other more privileged explanatory variables and theoretical approaches.

American and Japanese Interest in Radwaste Disposal

As was noted in chapter 2, Britain was dumping low-level radioactive waste in the Atlantic almost annually from 1949 on. Starting in 1967, waste delivered by Britain, the Netherlands, Belgium, and Switzerland was periodically dumped in a deep part of the Atlantic under a voluntary system of international supervision administered by the Nuclear Energy Agency. Although these dumping operations complied with the global ocean dumping regime when it went into force in 1975, several nondumping nations were opposed to them. For example, in 1978, at the consultative meeting of the members of the global dumping regime, "many delegations expressed the view that radioactive waste disposal in the ocean should be discouraged" ("Report of the Third Consultative Meeting," LDC III/12 (October 24, 1978), p. 16). As had happened in 1958 at the United Nations Conference on the Law of the Sea, the conflict surrounding radwaste disposal went unresolved.[2]

But international worries about radwaste disposal began to increase conspicuously in the late 1970s. In 1979 Japan announced that it planned to experimentally dump radwaste at a site north of the Mariana Islands in the South Pacific ("Japan Plans to Begin Ocean Disposal," *Nuclear News*, November 1980, p. 20). Japan, whose geography lacks sufficient long-term geologic stability for repository sites, shipped spent fuel to Europe for reprocessing at Sellafield, Britain, or at Le Hague, France.[3] According to the Japanese government, the plan conformed to the regulations of the global ocean dumping regime. Under the Japanese plan, 5000–10,000 drums of nuclear waste would be dumped in 1981, when Japan would become member of the global dumping regime. The dumping would be expanded to up to 100,000 curies a year after the Japanese government had verified the safety of its experimental program.

The Japanese plans caused an uproar in the Pacific region, however. A mission from the Commonwealth of the Northern Mariana Islands soon presented a formal anti-dumping petition, representing seventy organizations with a total of more than 12 million members, to the Japanese Diet.[4] The Japanese government was also asked to send officials to the islands for discussions. Encouraged by the United States, Japanese scientists and politicians toured the region, trying to persuade governments that the dumping was safe. The campaign failed, however, and the Foreign Minister of New Guinea subsequently told the UN General Assembly that the long-term effects of dumping could be catastrophic (Callick 1980, p. 40). In October of 1980, the governors of Hawaii, Guam, American Samoa, and the Northern Mariana Islands issued a statement opposing the dumping of radioactive waste planned by Japan and the United States and declared that their organization, the Pacific Basin Development Council, "totally opposes the dumping of radioactive nuclear waste in any part of the Pacific Basin" (Trumbull 1980).[5]

In February of 1981 the Japanese government indicated that, in response to the protests, it had put off its plans to begin experimental dumping later the same year (Kamm 1981).[6] A complex combination of government considerations and domestic opposition apparently led to that decision.[7] Instead, Japan hoped to allay the fears of the Oceanian and Southeast Asian countries and then to dump small amounts of radioactive waste; only if that proved safe (as was expected) would full-scale dumping begin, probably in 1987 or later (Ishihara 1982, pp. 469–470). In August of 1984, however, officials of the Japanese Science and Technology Agency reportedly admitted that they were prepared to break their 1980 assurance not to dump without the consent of the Pacific nations (Dibblin 1985a, pp. 18–19).[8] After yet another uproar, representatives of the same agency assured that dumping would happen only "with the understanding" of the Pacific nations (Dibblin 1985b, p. 21).

There also existed a real possibility that the United States, after a pause of almost 20 years, would resume ocean dumping of some forms of radioactive waste. Toward the end of the 1970s, the United States' regulation of ocean dumping—the essentially precautionary and prohibitory Marine Protection, Research, and Sanctuaries Act of 1972, examined in chapters 4 and 5—came under some attack. There were indications that existent stringent ocean dumping regulation would be relaxed as Congress and the courts became painfully aware of some of the realities of strict regulation, namely considerable pollution control costs and insufficient alternatives.

At the same time it was alleged that parts of the oceans had a considerable capacity to assimilate some wastes (Walsh 1981).[9]

Around the same time, the U.S. Environmental Protection Agency began to look to the oceans as a possible disposal alternative for both low-level and high-level radioactive wastes. Public concern over disposal of radioactive waste made it extremely difficult, if not impossible, to find sufficient permanent disposal facilities on land (Shabecoff 1982). One EPA official explained in 1981 (Dyer 1981, p. 11): "With increasing public concern for waste management practices on land and the need to find permanent disposal sites, the United States is again looking towards the oceans as a possible alternative to land disposal for both low-level and high-level radioactive waste."[10] In addition, when weighing the costs and benefits of regulation, as the Reagan administration urged the EPA to do, ocean disposal of old nuclear submarines was clearly more attractive than land disposal.[11] There also seemed to be a growing consensus among marine scientists that radwaste disposal would cause no significant risks to either human health or the marine environment.[12] Thus, in 1980 the EPA began revising existing regulations so that thousands of tons of slightly contaminated soil left over from the Manhattan Project and more than 100 retired nuclear submarines, each representing more than 50,000 curies of radioactive waste, could be dumped at sea.[13]

The possibility of a change in the EPA's policy on disposal of radioactive materials in the ocean sparked considerable alarm within the environmental community. Public attention was attracted to radioactive waste dumped from the 1946 to 1970 when drums were found unexpectedly and previously unknown dump sites were disclosed. The EPA sent some of its researchers to examine a former site near Boston for health effects.[14] As the EPA scientists expected, however, no evidence of harm was turned up.[15] The public visibility of the issue was further heightened as environmental groups organized public policy forums and "citizen workshops" that addressed past dumping in the United States, legal aspects of international regulation, and plans to bury high-level radioactive waste in the deep seabed. Hearings on the early dumping were held in San Francisco and in Boston, and a hearing on the proposed dumping of decommissioned submarines was held in North Carolina.[16] Environmental pressure groups, conservationist groups, private citizens, local business leaders, and commercial fishermen's organizations all advocated a ban on disposal of radioactive waste in the ocean.

The Oceanic Society, a Washington-based environmental group, challenged the scientific basis of the Navy's proposal and coordinated domestic opposition to ocean dumping of radioactive waste.[17] Governor Thomas Kean of New Jersey called the proposal to dump off the New Jersey coast "a very severe potential health hazard" ("Kean Assails Proposal On Dumping A-Waste," *New York Times*, March 28, 1982).

In September of 1982, to head off the Reagan administration's proposal, the House of Representatives approved legislation imposing a two-year moratorium on any dumping of low-level radioactive waste ("House Backs Moratorium on Ocean Dumping," *New York Times*, September 21, 1982).[18] Despite assurances by ocean scientists and experts that the risks were minuscule, the practice was perceived as a threat to the marine environment. "As a common-access resource, the ocean is not protected by the same economic and political forces that protect private property," said one of the bill's sponsors. "It is up to the members of Congress to provide a voice for the ocean and to insure that the ocean has sufficient protection. We are specifically charged with the mandate of providing our citizens and our future generations a healthy and unpolluted ocean environment." (Rep. Norman E. D'Amours (D, New Hampshire), chairman of the Subcommittee on Oceanography, *Congressional Record: House*, September 20, 1982, p. H 7261) The Senate approved the bill in December of 1982, although some powerful senators and President Reagan were opposed to it (*Congressional Quarterly: Weekly Report*, December 25, 1982, p. 3138).[19] The sponsors in the House outmaneuvered them, however, by attaching the bill to a gas-tax bill supported by the Senate and the president (source: interview with anonymous U.S. government source). In 1984, the U.S. Navy made it official that it had decided to bury the defueled radioactive engine compartments of its retired submarines on government-owned land.[20]

In summary, unlike the early 1970s, when a few European governments (Britain, Switzerland, the Netherlands, Belgium) were dumping low-level radioactive waste, the United States and Japan planned to resume ocean dumping in the 1980s. Since the mid 1970s, several nuclear nations had furthermore been examining the technical and scientific feasibility of high-level radioactive waste disposal into the deep ocean seabed.[21] Thus the regulatory situation had changed dramatically since the global ocean dumping regime had been created. Originally, the United States strictly regulated low-level radioactive waste disposal. And, as was noted in chapter 5, the congressional sponsors of the bill to regulate ocean dumping had been strongly opposed to the dumping of "hot" high-level radioactive waste at sea.

Greenpeace's Campaign against Radwaste Disposal

In 1978, Greenpeace launched a campaign against the European dumping in the Atlantic Ocean.[22] Greenpeace, since its founding in Canada in 1971, had become arguably the world's largest and most effective international environmental pressure group. The European dumping operation took place at a site 4000 meters below sea level, located approximately 700 kilometers off Spain's northwest coast. The nuclear industry feared that the campaign also was intended to mobilize public opinion to prevent a resumption of dumping by the United States or a start of such operations by Japan (Rippon 1983, p. 79).

Greenpeace intended to hinder the annual European dumping operation. The organization hoped to obstruct the dumping operation by positioning inflatable dinghies underneath the dumping ship's platforms from which the containers were rolled into the sea. The dinghies were operated from the *Rainbow Warrior*—the same ship was used to protest against French nuclear tests on the Polynesian island of Mururoa in the South Pacific.[23] The Greenpeace ship followed a freighter to the dump site. At a press conference in Britain, Greenpeace showed film of its unsuccessful attempts to keep the dumper ship from tipping more than 5000 barrels of radioactive waste into the sea. Greenpeace's charge that the dumping violated the rules of the global dumping regime was rejected by the British government, which declared that the material dumped only had insignificant amounts of radioactivity (Morris 1978).[24] In the summer of 1979, Greenpeace again set out to obstruct the dumping. The British press reported how the dumper ship crew's used powerful fire hoses to prevent the protesters from placing dinghies under the cranes dumping drums of radioactive waste into the sea (Allen-Mills 1979).

In 1982, three Dutch ships were responsible for the annual dumping operation. The international press reported how dinghies launched from the *Rainbow Warrior* were placed under the ships' cranes. People on Spain's northwest coast, where fishing is one of the main industries, strongly protested against the dumping and dispatched two trawlers with local politicians on board to escort the Greenpeace ship to the dumping site. Dumping was suspended 24 hours after two drums struck a dinghy carrying Greenpeace protesters.[25] Spanish politicians' appeals by loudspeakers to the captain also interrupted the operation, which was resumed but then canceled.[26] Local politicians and thousands of people carrying anti-dumping posters welcomed the Greenpeace crew when it arrived in port in Vigo in Galicia after having protested against the dumping, and the Socialist government headed by Prime Minister Felipe Gonzalez, which came to power in 1982, vowed to stop the dumping taking place off the Spanish coast (interview, José Juste Ruiz, November 29, 1991). After this incident, the European Parliament adopted a resolution urging the European Commission to employ "any procedures at its disposal, either action within the Community framework, or through international agreements, to stop this dumping of nuclear waste."[27] At this point, "the issue of the annual dump was developed into an international scandal" by Greenpeace (Pearce 1991, p. 54).

Greenpeace's campaign was also highly successful elsewhere in Europe. Within a month after the aforementioned incident, the Dutch government officially suspended all dumping of radioactive waste. "This ministry is convinced that ocean-dumping is a safe disposal for wastes," said a spokesman for the Dutch Ministry of Public Health and Environment, "but it's clear that our society does not want ocean-dumping."[28] Transport and loading of waste had been possible in the previous two years only because police had thwarted protests organized by Greenpeace.[29] Considerable public protest at a time when Dutch politics was increasingly becoming "green" caused the government to reverse its policy (source: interview with Dutch delegate to 1991 London Convention consultative meeting, London, November 27, 1991).[30] In 1980, there were also protests in Zeebrugge, Belgium, where demonstrators caused considerable damage to instruments on the bridge of a freighter leaving to dump in the Atlantic ("Demonstrations Against Low-Level Sea Dumping," *Nuclear News*, August 1980, pp. 72–73). In 1982, national attention was again attracted to the issue when an attempt by the mayor of Bruges to prevent shipment of waste destined for the ocean through his city was overruled by the Belgium government (Bourke 1983, p. 25).

In this atmosphere of changing public opinion, governments increasingly felt pressured to cancel dumping. As illustrated by a leading article appearing in the London *Times* on the eve of that year's consultative meeting, British mass media disapproved continuing radwaste disposal in 1983: "In the long run, sea dumping is not a desirable practice. It is in principle a bad idea to put things that may be dangerous where you cannot keep an eye on them. Too little is known of the sea bed, underwater currents and the food chains of marine life for the sea to be suitable for use as an oubliette on an indefinitely expanding scale." ("Deep-Sea Dumping," London *Times*, February 16, 1983) Radwaste disposal was seen as a dangerous activity and scientific knowledge was perceived as being too uncertain to guide policy. Public opinion in Europe was increasingly questioning the wisdom of existing policy.

The 1983 Radwaste Disposal Moratorium

Encouraged by the success of the Greenpeace campaign, the anti-dumping governmental opposition launched an attack on the scientific basis of the radwaste disposal policy and on regime rules at the 1983 consultative meeting. Banning radwaste disposal required that the gray and black lists of the London Convention be amended. In accordance with the London Convention, low-level radioactive waste would have to be moved from the gray list to the black list. The convention stipulated, in addition, that any amendment to the black and gray lists "will be based on scientific or technical considerations."[31] To halt radwaste disposal, it would therefore be necessary to present scientific and technical evidence that such practice was harmful and should be banned under the convention.[32] But the regime's strong emphasis on scientific evidence and proof of environmental damage did not mean that scientific and technical knowledge shaped regime development or facilitated policy coordination as the knowledge-based regime approach claims.

Greenpeace's campaign and developments in the Pacific expanded the governmental coalition against radwaste disposal. By 1983, two Pacific islands, Kiribati and Nauru, had become members of the global ocean dumping regime in the hope that the convention could be amended to ban all forms of radioactive waste disposal at sea (van Dyke et al. 1984, p. 743).[33] Nauru, represented by an American anti-nuclear campaigner—Jackson Davis, a biology professor from the University of California—proposed an immediate global ban on radwaste disposal.[34] Being heavily dependent on marine resources, fish being one of the two staple foods and an important economic resource, Kiribati and Nauru feared that radioactive waste endangered the marine environment and presented a scientific report in support of their claim. Their report claimed that radioactivity had leaked from old drums into the marine environment and had entered into the oceanic food chain, that existing knowledge of behavior of radioactivity in the ocean was based on incorrect and uncertain theoretical models, and that experts disagreed on low-level radiation hazards.[35]

The five Nordic countries—Denmark, Finland, Iceland, Norway, and Sweden—proposed a ban on dumping to start in 1990. They agreed in principle with the proposal of Kiribati and Nauru but wished to give dumper nations enough time to develop land-based alternatives. In the intermediate period, dumping should be more strictly controlled and the amount of waste should not exceed the present level. Furthermore, only existing dump sites should be used, and no new dumpers should be allowed. The Marine Pollution Division of Denmark's National Agency of Environmental Protection (NAEP) formulated the Danish policy, but Denmark failed to win Nordic support for the Nauru and Kiribati proposal. Danish scientists, who like their international peers considered the risks of dumping low-level radioactive waste to be very low, were not conferred with (interview, Asker Aarkrog, March 20, 1992). Two of the Danish government officials later joined Greenpeace (interview, Kirsten F. Hansen, January 17, 1990).

The Spanish delegation told the consultative meeting that dumping in the North Atlantic was a cause of great domestic public concern. Spain considered that the effects on human health and long-term consequences of dumping were the subject of scientific controversy and proposed suspension of dumping operations until the necessary research and evaluation were completed. The delegation from Ireland, one of the countries nearest the dump site then in use, was opposed in principle to the dumping of radioactive waste at sea and supported the Nordic proposal. The Irish government was "coming under increasing domestic pressure from a public opinion which was not convinced that dumping did not constitute a hazard" (LDC 1983a, pp. 22–23). Ireland maintained that governments wishing to dump had the responsibility to demonstrate that dumping was safe.

But pro-dumping states strongly defended their policy and the scientific basis of radwaste disposal. Britain replied that the documents submitted by Kiribati and Nauru did not provide the scientific and technical basis required for amendment of the convention. The convention should consequently not be amended. The British delegation was of the opinion that the onus of proof that dumping was unsafe rested with those proposing to change the convention. Britain failed, however, to get support for this view (Edwards 1983, p. 6).

Switzerland fully supported the British position. The United States supported the British position too, stressing that a change of the convention to ban radwaste disposal should be based on sound scientific evidence of adverse health effects and damages to the marine environment. A scientific advisor, Charles D. Hollister of the Woods Hole Oceanographic Institute, one of America's most respected marine research centers, concluded that the Nauru report "is clearly not the balanced scientific evaluation claimed by the authors and thus it is my recommendation that no amendments to the London Dumping Convention be considered until such an evaluation is completed" (LDC 1983b, Annex 1).[36]

Why did the United States not support a global nuclear dumping ban? As mentioned already, the United States introduced a domestic radwaste disposal moratorium in 1982. Realists and probably also epistemic-community theorists suspect that the United States supported the regime on this issue because it paralleled the recent policy change in the United States. Importantly, however, the U.S. administration did not welcome the legislation on radwaste disposal passed by Congress in 1982. The U.S. Navy was still faced with the problem of disposing of its retired nuclear submarines and preferred to keep the option of ocean disposal open. Moreover, the U.S. marine scientific community generally did not support an unqualified ban on ocean dumping of wastes, radioactive wastes included; U.S. legislators and the public, scientists believed, exaggerated the risks involved in ocean dumping. A report released in 1984 by the National Advisory Committee on Oceans and Atmosphere, co-written with the National Oceanic and Atmospheric Administration, indirectly recommended that Congress and the administration revise the policy of excluding the use of the ocean for low-level radioactive waste disposal.[37] In the view of the U.S. marine scientific community, an international ban on nuclear ocean dumping would instead be similar to "doing the same mistake twice" (interviews, Bryan C. Wood-Thomas, August 29 and November 27, 1991). Environmental concerns did not cause the administration to reverse its pro-dumping foreign policy; hence, foreign policy (the domain of the executive branch of government) differed from domestic policy. Hegemonic leadership was absent, but, despite the realist claim to the contrary, this did not mean that regime rules, principles, and norms had little influence on states or that states could pursue their individual policy goals.

The Netherlands delegation explained to the meeting that it was looking for possibilities to avoid dumping from 1983 and intended to store waste on land. Owing to difficulties in finding suitable disposal alternatives, dumping in 1983 could perhaps not be avoided. Japan believed that sea disposal of radioactive waste would not adversely affect the marine environment when international regulations, which presently rested on firm scientific basis, were followed. The Japanese government therefore strongly opposed proposals for prohibiting sea disposal.

During informal negotiations among the various delegations, it became clear that the proposal to amend the convention would not receive support from a sufficient number of governments. Agreement was reached, however, that the scientific basis of the proposal by Nauru and Kiribati should be reviewed by an expert group. The results of such a study should be discussed in 1985, at which time further action should be taken.

The transnational opposition's attack on the scientific basis of radwaste disposal thus failed, but the regime provided anti-dumping nations with other ways of protecting their interests. Spain proposed a moratorium resolution (according to the London Convention, resolutions require a simple majority) that meant a suspension of all dumping at sea pending completion of such an expert group study of effects of dumping of low-level radioactive waste on the marine environment and human health (LDC 1983a, Annex 3: Resolution LDC.14(7) Disposal of Radio-Active Wastes and Other Radio-Active Matter at Sea). In a subsequent roll call vote, which the United States and Britain failed to block, 19 countries—Spain, Portugal, the Nordic countries, Ireland, Canada, and almost all developing countries—voted in favor of the Spanish proposal (Curtis 1983a,b).[38] (See table 8.1.) The sponsors of the moratorium resolution easily persuaded the developing countries[39] to support the moratorium (interview, José Juste Ruiz, November 29, 1991). The group of countries considering or involved in dumping voted against the resolution. Five countries abstained. Although the moratorium resolution was not legally binding on governments, several delegations indicated that it was morally binding. The nuclear industry, among others, thus expected that continued ocean dumping would

Table 8.1
Votes on the 1983 Low-Level Radioactive Waste Moratorium. Source: LDC 1983a, p. 29.

In favor (19)
Argentina, Canada, Chile, Denmark, Finland, Iceland, Ireland, Kiribati, Mexico, Morocco, Nauru, New Zealand, Nigeria, Norway, Papua New Guinea, Philippines, Portugal, Spain, Sweden
Against (6)
Britain, Japan, Netherlands, South Africa, Switzerland, United States
Abstaining
Brazil, Federal Republic of Germany, France, Greece, Soviet Union

"result in a substantial political storm" ("A Call for a Two-Year Halt on Ocean Disposal," *Nuclear News*, March 1983, p. 120).

Contradicting realist regime analysis, the hegemon and powerful states could not determine a regime development that diverged significantly from their interests. Britain immediately indicated it would not be bound by the decision (Wright 1983b). Britain planned to dump 3500 metric tons of low-level radioactive waste, representing more than 1500 curies of alpha radiation and some 150,000 curies of beta and gamma radiation, in the Atlantic (Pearce 1983, p. 924). The Swiss delegation also expressed the view that Switzerland did not feel bound by the resolution. Switzerland and Belgium intended to dispose of relatively small amounts later that year, but Switzerland would stop dumping in 1984 (Edwards 1983, p. 6). The government of the Netherlands explained that it had difficulties disposing of low-level radioactive waste on land and therefore might have to carry out dumping in the summer 1983. Later it became clear that the French government, which had not participated since 1969, also intended to participate in the 1984 dumping ("Ocean Disposal Operations to Continue," *Nuclear News*, July 1983, p. 50).

To prevent the scheduled dumping, Greenpeace set out to strengthen and broaden the transnational anti-dumping coalition by including a significant number of stakeholders and special-interest groups. Greenpeace made contact with the National Union of Seamen, the British seamen's organization, hoping that the union would boycott the dumping planned for the summer of 1983 (Pearce 1991, pp. 54–55).[40] The initiative was successful. In March of 1983, the British seamen, concerned primarily about their safety when handling the waste,[41] passed a resolution in favor of halting ocean dumping of radioactive materials ("Four Unions Back Ban on A-Waste Dumping," *Guardian*, April 7, 1983).[42] A month later, the opposition was further strengthened when the Transport and General Workers' Union, the train drivers' union known as ASLEF, and the National Union of Railwaymen, at a meeting organized by Greenpeace, agreed on an attempt to halt ocean dumping of radioactive waste.[43] In June of the same year, the British seamen announced a ban on handling the waste. The seamen refused to crew a "Greenpeace-proof" ship (fitted with a hole in the hull through which drums of waste could be dropped) that had been chartered by Britain, Belgium, and Switzerland to carry out dumping (Pearce 1983).[44] The Transport and General Workers' Union and ASLEF similarly called on their members not to handle or transport the waste. Transport union boycotts were also called in Switzerland and in Belgium (Curtis 1984, p. 68). If the British government let the armed forces carry out the dumping, as the unions expected, an armada of protest vessels was expected to sail from Spain to converge on the dumping site. "We understand there are already plans for quite a lot of vessels to leave Spain," explained an executive officer of the Transport and General Workers' Union, "and we would hopefully form part of that armada" (Ardill 1983). In February and July of 1983, Spanish Friends of the Earth, ecologists, and left-wing protesters demonstrated before the British Embassy in Madrid in protest against the plan to dump. In July, more than 150 British flags were burned in several towns and cities in Galicia, and in one city Prime Minister Margaret Thatcher was burned in effigy.[45]

Mounting national and international pressure orchestrated by Greenpeace forced the British government to reverse its policy. In September, the British opposition was further strengthened when the seamen's union won backing from the Trade Unions Congress for a motion condemning the use of the world's oceans as a dumping ground for nuclear waste and demanding that development of land-based disposal facilities be accelerated. The position of the seamen was that "radioactive waste should not be dumped irretrievably but should be stored in above-ground, engineered facilities in a location acceptable to the local communities involved."[46] The Trade Unions Congress furthermore urged the British government to comply with the decision made at the February meeting of the global ocean dumping regime; this decision was carried by a vote of

7,150,000 to 2,764,000.⁴⁷ At the end of August 1983, it was reported that the British government had given up its dumping plans, together with the Belgian and the Swiss governments.⁴⁸ On the eve of the 1985 meeting of the members of the global dumping regime, the general secretary of the Trade Unions Congress and the British seamen reiterated their opposition to any British plans to resume dumping.⁴⁹ The International Transport Unions Federation, in addition, was "putting its full weight behind a ban and could force dumping nations to toe the line" (Dibblin 1985b, p. 21). An ENGO had initiated a regime development which scientific consensus did not support and which the hegemon and powerful states could not prevent.

The 1985 London Convention Resolution

Within the next few years the transnational anti-dumping coalition further undermined both the historical dominance of power politics in this issue area and the scientific justification for radwaste disposal. The eighth consultative meeting of the global dumping regime, in February of 1984, agreed on a more precise structuring of the review of effects of dumping of low-level radioactive waste on the marine environment and human health.⁵⁰ It was decided that a panel of international experts nominated by the International Council of Scientific Unions (ICSU), an UN-based advisory scientific body, and the IAEA should prepare a basic document which later would be examined by an expanded panel including experts from governments, international organizations, and ENGOs. This decision was a compromise between a group led by Britain, which wanted the IAEA and the ICSU to select the experts to review the evidence and make recommendations for consideration at the next consultative meeting, and another group of governments, led by Canada and Nauru, which felt that experts reflecting different interests and regions should review the evidence and make recommendations.⁵¹ The United States in particular insisted that the representatives from NGOs were indeed experts in the relevant fields.⁵² Experts should be knowledgeable in fields such as marine ecology, oceanography, radiological protection, marine geochemistry, and marine mathematical modeling.⁵³

In their final report, completed in the spring of 1985, the experts did not make a recommendation on whether to amend the London Convention. They judged that the question was not a wholly scientific-technical one, and that "such a decision could involve value judgments which go beyond consideration of the technical and scientific evidence."[54] However, as to the risks of dumping, the experts concluded that "the calculations show that any risk to individuals from the use of the [Atlantic] dump site is very low, both in relation to other common radiation risks such as that from natural background radiation and to the risk that corresponds to any of the dose limits or upper bounds that would apply following current international radiation protection recommendations" (LDC 1985d, p. 136).[55] In other words, no scientific consensus existed to support and justify a halt on nuclear ocean dumping. The coming consultative meeting would have to reconsider the moratorium without clear recommendations from its scientific advisers.

Within the expanded group of experts, including representatives from governments and Greenpeace, some representatives proposed to make a clear statement that could be used by the consultative meeting in reaching a final decision.[56] They suggested that "no scientific or technical grounds could be found to prohibit the dumping at sea of all radioactive wastes, provided that dumping is carried out in accordance with internationally agreed procedures and controls" (LDC 1985a, p. 26). But a number of representatives opposed any such categorical statement. There was agreement on a compromise stating "no scientific or technical grounds could be found to treat the option of sea dumping differently from other available options when applying internationally accepted principles of radioprotection to radioactive waste disposal."[57] The British press, however, reported that "all the parties who attended seem to come away with a different version of the result" (Brown 1985a).

Ignoring expert opinion, the transnational anti-dumping coalition continued to attack the scientific and conceptual basis of the radwaste policy. The report of the expanded panel was the focus of the ninth consultative meeting of the global dumping regime, held in September of 1985.[58] Governments reached very different conclusions from the findings of the report. Nauru, Spain, Denmark, Norway, Australia, New Zealand, Saint Lucia, Iceland, and Brazil found the report supported their fears about radioactive dumping. Several of them stressed that land disposal was safer, more controllable, and more reversible than ocean disposal. Governments that were in the process of developing land-based alternatives (Finland, Sweden, the Netherlands, and the Federal Republic of Germany) also opposed ocean dumping. Spain and Ireland explained that factors other than scientific and technical ones—for example, availability of land-based disposal alternatives—also should be taken into account. Several of the governments opposing dumping and several international environmental organizations stressed that available knowledge was insufficient for it to be modeled adequately and with a sufficient margin of safety. A representative of the scientific panel, however, objected that the scientific findings were being "ignored, distorted or misinterpreted by some parties in unprofessional attempts to exaggerate the uncertainties in that report" (LDC 1985c, p. 22).

Japan explained that, although it presently did not intent to dump without the consent of the Pacific region, it needed to dispose of radioactive waste, and that, as a small country, it had to consider ocean disposal. Provided that scientific and technical studies showed disposal would be safe, the option should remain open. France concluded that no scientific grounds for suspension of ocean dumping had been found, and that the option should be reopened. Britain and the United States also argued that the available scientific evidence did not support a change of the convention. The United States suggested ending the suspension of dumping. Belgium and Switzerland supported the position of the United States. Thus, the panel report did not help to resolve the conflict. Later, one member of the U.S. delegation remarked that "both those for and those against sea disposal have pointed to the panel's conclusion as vindication of their own positions" (Sielen 1988, p. 10). "Scientifically," Sielen later noted in an interview (August 29, 1991), "you seem to be able to argue either way."

Intense negotiations followed but did not result in agreement. Although Britain had hoped to avoid a vote altogether, a resolution co-sponsored by Spain and 15 other states for an indefinite moratorium pending further considerations of the issues involved was then brought to a vote (Brown 1985b). The group of governments supporting a moratorium had grown (mostly because several developing countries had joined) to 26; five governments (almost the same ones that had been against the 1983 moratorium) opposed it; and seven abstained (see table 8.2).[59]

Governments opposing the moratorium resolution protested fiercely against the vote. The British press reported "UK threatens to withdraw from convention on nuclear dumping" and "the big nuclear nations, including the United States, had pointed out that they would have to recon-

Table 8.2
Votes on the 1985 Low-Level Radioactive Waste Moratorium. Source: LDC 1985a, p. 37.

In favor (26) Australia, Brazil, Canada, Chile, Cuba, Denmark, Dominican Republic, Finland, Germany, Haiti, Honduras, Iceland, Ireland, Kiribati, Mexico, Nauru, Netherlands, New Zealand, Norway, Oman, Panama, Papua New Guinea, Philippines, Spain, St. Lucia, Sweden
Against (5) Britain, France, South Africa, Switzerland, United States
Abstentions (7) Argentina, Belgium, Greece, Italy, Japan, Portugal, Soviet Union

sider their position if dumping was banned" (Brown 1985c). The United States cautioned governments that "similar action in the future on other important issues will not only undermine the fabric and regulatory framework of the London Dumping Convention, but also tend toward its politicization."[60] The Canadian chairman of the consultative meeting again appealed for a compromise at a subsequent press conference, saying that it was "better for all countries to take one step forward than some to take five steps back and others none at all" (Brown 1985c).[61]

The resolution called for suspension of all ocean dumping of radioactive waste pending studies of wider legal, social, economic and political aspects of resuming radwaste disposal (LDC 1985c, Annex 4: Resolution LDC.21(9) "Dumping of Radioactive Wastes at Sea"). Thus the resolution was intended to broaden the regime's decision-making principle to include considerations other than scientific and technical ones. While the exact nature of such considerations were not clearly spelled out, future proposals to dump should be examined in the light of what international law said about liability, duty to cooperate, and the oceans legal status as "common heritage of mankind." Economic considerations would, or could, include losses to the fishing industry. Dumping in the Atlantic, for example, has repeatedly resulted in reduced sales of fish in Spain.[62] In 1980, the Japanese market for sablefish collapsed after a photograph of a sablefish swimming near drums dumped in the Pacific off San Francisco was published in newspapers around the world. All orders for sablefish, not just ones from the West Coast of the United States, were canceled (Bishop 1991). Risks and costs of land disposal also had to be examined.

An examination would have to be made of whether it could be proven that radwaste disposal would not harm human health or cause significant damage to the marine environment. Crucially, the resolution thus shifted the onus of proof to those interested in dumping who in the future would have to demonstrate that no harm would be inflicted on the marine environment or humans. This decision, in particular, was a significant victory for those opposing radwaste disposal, and delegations considered that such a proof could not be made (Brown 1985b).[63] This regime development evidently did not reflect scientific opinion, as the knowledge-based approach would expect.

At the tenth consultative meeting of the global ocean dumping regime, held in October of 1986, it was decided that scientific and technical aspects of radwaste disposal would be examined by one group under an Intergovernmental Panel of Experts on Radioactive Waste Disposal at Sea (called IGPRAD), while other groups would consider legal and social aspects of resuming dumping.[64]

In 1988, the British government formally announced that radwaste disposal would not resume (Flowers 1989, p. 108). Significantly, Britain continued to reserve a right to resume radwaste disposal of some large waste objects.[65] In 1989, however, a proposal to dump decommissioned nuclear submarines by the British Ministry of Defence was rejected by ministers.[66] While domestic regulation in the United States practically prohibited ocean dumping of low-level radioactive waste, the administration had not definitively canceled plans to dump low-level radioactive waste.[67] But resuming dumping, which would need the approval of Congress, was unlikely given public sentiment on this issue (interview, Robert S. Dyer, September 27, 1991).[68] Japan had no plans to dump, although domestic law did not rule out the option (interview, Takao Kuramochi, August 30, 1991).

An attempt to amend the London Convention to finally prohibit dumping of low-level radioactive waste surfaced at the fourteenth consultative regime meeting, held in November of 1991.[69] But the meeting wished to postpone the decision on such a conference until after the so-called Rio Conference (officially the United Nations Conference on Environment and Development) in June of 1992. In 1993, after a series of meetings for which governments, and later also ENGOs, had prepared studies and papers, IGPRAD presented the London Convention meeting with options for deciding on radwaste disposal.

The 1993 Radwaste Disposal Ban: Emphasis on Precautions

In the late 1980s, Greenpeace, Scandinavian and certain Northern European countries, and a few marine scientists successfully advocated the precautionary principle at the scientific working group meetings of the global ocean dumping regime.[70] They succeeded in changing the underlying principles and norms of the regime as well as the scientific basis of regulation, with significant consequences for the radwaste disposal issue.

According to the precautionary principle, unanimously adopted by the 1991 annual London Convention meeting, "preventive measures are taken when there is reason to believe that substances or energy introduced in the marine environment are likely to cause harm even when there is no conclusive evidence to prove a causal relation between inputs and their effects" (LDC 1991, Annex 2). Equally significant, the concept of the assimilative capacity was rejected as the scientific principle underlying ocean dumping regulation, and countries agreed that "existing pollution control approaches ... have been strengthened by shifting the emphasis from a system of controlled dumping based on assumptions of the assimilative capacity of the oceans, to approaches based on precaution and prevention" (ibid.). By this decision, governments shifted the emphasis from "dispose and dilute" approaches to "isolate and contain" approaches. Countries had earlier often been deadlocked because some, especially Britain, traditionally preferred regulation based on the assimilative capacity of the oceans, while others followed a more cautious approach.[71] Also the scientific debate on the radwaste disposal issue to a significant degree revolved around the concept of the assimilative capacity.[72]

In July of 1993, IGPRAD finalized its work. The final report listed seven policy options, but IGPRAD did not recommend any in particular as this would have been outside the terms of reference for its work. However, the report noted "the growing awareness within the national and international communities that new and more effective measures are needed to protect the global marine environment" (LC/IGPRAD 1993, Annex 2, p. 50). At the legal level, during the past 20 years "a trend towards, first, restricting and controlling, second, prohibiting sea disposal of radioactive wastes on a regional basis" was acknowledged in the report (ibid.). In regard to the scientific and technical issues, it was noted that ocean disposal, in comparison with other disposal alternatives for radioactive waste, could result in transboundary transfer of radioactive materials and relative difficulties in monitoring and retrieval of ocean dumped radioactive waste packages. However, in what was obviously a compromise solution, it was also concluded that "the same internationally accepted principles of radiological protection apply equally to the scientific and technical assessment of all radioactive waste disposal options" (LC/IGPRAD 1993, Annex 2, p. 51).

Rather incompatible with knowledge-based analysis, scientific consensus still did not lead to a change in anti-dumping governments' perception of the radwaste disposal issue and, judging by its initial ambition to create a consensus-building process, the IGPRAD process was not a success.[73] Some members were of the opinion that the risks from dumping at the Northeast Atlantic dump site were "considerably smaller than the risks associated with exposures to naturally occurring radionuclides and certain organic chemicals in seafoods" (LC/IGPRAD 1993, Annex 2, p. 47). Other members emphasized instead the uncertainty of theoretical models and lack of knowledge of essential issues. Similarly, some believed that ocean disposal should be included when conducting comparative risk assessments, whereas others wished to exclude ocean disposal *a priori* because of the transboundary nature of radwaste disposal and problems in monitoring and retrieval of radioactive waste dumped.

An event of great political significance occurred in October of 1993, when Greenpeace exposed a Russian warship dumping nearly 900 metric tons of liquid low-level radioactive waste into the Sea of Japan (Hiatt 1993). Japan had not been informed about the dumping. The Russian navy subsequently explained that Russia did not have the land capacity to store waste produced by the Russian nuclear-powered fleet.[74] The dumping caused strong concern in Japan as it took place only a few days after President Boris Yeltsin had been on a visit to Japan and the two countries had signed an agreement to end nuclear contamination of the oceans. Also, it sparked widespread concern in Japan about the possible contamination of fish and other sea life. Responding to protests from Japan, South Korea, the United States, and other countries, Russia canceled plans to dump another cargo of 700 metric tons of radwaste into the Sea of Japan. Immediately after the incident, Japan announced it would support a nuclear dumping ban at the 1993 London Convention meeting.[75]

In the fall of 1993, after an interdepartmental power struggle, a significant reversal of U.S. foreign policy on radwaste disposal took place. In early

November, the Clinton administration announced that it had decided to press for a legally binding worldwide ban on the dumping of low-level radioactive waste at sea, a radical departure from the policy of previous administrations (Pitt 1993b). The decision was taken after the issue of radwaste disposal had received prominent coverage in the media and after lobbying by Greenpeace and politicians. The Department of Defense unsuccessfully opposed this decision, which it felt interfered with vital interests of the Navy (Pitt 1993a). Globally, the balance was shifting in favor of a nuclear dumping ban despite the fact that Britain, France, Japan, and the United States lacked sufficient permanent land-based waste disposal facilities for their low-level radioactive waste.

In July of 1993, Denmark announced that it would call for formal action at the upcoming London Convention meeting on an amendment to permanently ban radwaste disposal. Denmark was supported in this by more than twenty other governmental delegations. In November, a number of governments submitted amendment proposals in regard to radwaste disposal to the 1993 London Convention meeting. A draft resolution was prepared by a working group and later adopted by vote (LC 1993, Resolution LC.51(16)). Thirty-seven countries voted for, none against; five (Britain, France, Belgium, the Russian Federation, and China) abstained (see table 8.3).

Belgium explained later that, owing to its small size and its population density, it did not wish to exclude any alternative solution to land disposal. China explained that, since studies and assessments carried out by IAEA had not been completed, the ban could not take these results into account. China was not, however, in favor of ocean dumping. France objected to the

Table 8.3
Votes on the 1993 Permanent Low-Level Radioactive Waste Ban. Source: LC 1993, pp. 16–17.

In favor (37)
Argentina, Australia, Brazil, Canada, Chile, Cyprus, Denmark, Egypt, Finland, Germany, Greece, Iceland, Ireland, Italy, Japan, Luxembourg, Malta, Mexico, Morocco, Nauru, Netherlands, New Zealand, Nigeria, Norway, Oman, Papua New Guinea, Philippines, Poland, Portugal, Solomon Islands, South Africa, Spain, Sweden, Switzerland, Ukraine, United States, Vanuatu

Against (0)

Abstentions (5)
Belgium, Britain, China, France, Russian Federation

ban in principle, arguing that it was not based on objective scientific grounds. The Russian Federation explained that, owing to a lack of sufficient land-based disposal facilities, it had hoped for a grace period until December 31, 1995. Since the London Convention meeting disagreed, the Russian Federation chose to abstain.

After the vote, abstaining countries found themselves in an embarrassing position, and all of them except the Russian Federation informed the London Convention secretariat soon after the meeting that they accepted the nuclear dumping ban (Castaing 1993).[76] The impact of international public opinion was very important when governments agreed to accept the ban. "The UK recognizes," the British Minister of Agriculture explained, "that the weight of international opinion on this matter means that such dumping is not, in any event, a practical proposition. We have, therefore, decided to accept the ban."[77] To save face, the Belgian government also accepted the resolution. Similarly, to end the controversy caused by its abstention, and pressured by environmental groups, the French government reversed its pro-dumping policy and decided to adopt the radwaste disposal ban.[78]

Summary

I have demonstrated that an ENGO acted as a catalyst for change within the regime and forced the transformation from regulation to prohibition despite scientists and powerful states supporting radwaste disposal. Greenpeace focused international public opinion on the issue of radwaste disposal, mobilized interest groups and stakeholders, pressured governments to halt dumping, and helped in changing the regime's principle in regard to what governments should do in the face of scientific uncertainty. ENGO pressure was strongly amplified by the global dumping regime which established norms and standards for acceptable behavior within this specific issue area. These norms and standards reflected international public opinion and the greening of civil society and governments; they did not reflect consensus within a scientific community, or protect the interests of powerful states. Because of the regime norms and principles, some policies were politically legitimate while others were not. States disregarding regime standards and norms were subject to domestic and international criticism and scorn, and even powerful states therefore chose to comply.

9

Explaining Regime Change

The Change of the Global Ocean Dumping Regime

The global ocean dumping regime changed in reasonably identifiable stages. In one early stage, Greenpeace focused international public opinion on radwaste disposal and forced the issue onto the international environmental agenda. Greenpeace was an adept practitioner of public diplomacy, and it used international public opinion as an effective political instrument in shaping and influencing regime development.

Greenpeace's campaign reflected widespread public concern about radwaste disposal. In the 1980s, supporters and opponents of radwaste disposal were in agreement that protests against dumping should be seen as manifestations of international public opinion. "Public opinion," one commentator noted in the journal *Nuclear Engineering International*, "can find expression in policy not only through appeals to government, but also through direct intervention. The seamen can be condemned for taking the law into their own hands, but their action is only a symptom of an underlying public concern which is apparent world-wide and which stochastic assurances of safety have done little to assuage." (Cruickshank 1983, pp. 13–14) In almost identical words, the pro-environmentalism journal *AMBIO* observed that "even though the moratorium was legally non-binding, trade unions in Britain and throughout the world heeded the message of international opinion" (Branch 1984, p. 330). According to the British press, the Spanish protests and demonstrations against the dumping planned for the summer 1983 had "strong British and international support" (Cemlyn-Jones 1983). And, as the Secretary General of the British seamen explained to the readers of the London *Times*, the seamen had international if not worldwide support: "The NUS has been inundated with messages of support from individuals and organizations around the world, including Jacques Cousteau, the mayors of towns and cities along the French and Spanish Atlantic seaboard, scientific groups, environmentalists and seafarers' unions.... It is ironic that it has taken a successful act of defiance against Government policy by three unions to protect Britain's good name in the international maritime community." (Slater 1983) With the 1983 moratorium resolution signaling the turning point, international public opinion no longer accepted ocean dumping of radioactive materials. In the eyes of the public, the previous policy had lost its legitimacy.

Greenpeace was also influential in the next stage of regime change. In the period 1983–1985, it pressured governments to refrain from radwaste dumping, and it mobilized stakeholders and interest groups. Greenpeace linked the domestic level and the regime level in a number of effective ways. By patrolling the oceans on the lookout for dumping ships, it created a sort of monitoring of and forced compliance with the regime. As was described in chapter 8, it instantly communicated information about the 1993 Russian nuclear dumping in the Sea of Japan to the Japanese government, which immediately reversed its policy on the issue. By exposing breaks of regime rules, Greenpeace increased the political costs of noncompliance by states. Moreover, Greenpeace coupled interest groups at the domestic level with those at the regime level. Through its recruitment of the British trade unions in 1983, it successfully forged an alliance with domestic actors in order to pressure a recalcitrant government to comply with international environmental rules. The British seamen had regime interests and urged the British government to comply with the rulings of the dumping regime. To build European opposition to radwaste disposal, Greenpeace also mobilized the Spanish opposition to nuclear ocean dumping. Opposition to radwaste disposal was increasingly transnationally organized, and protests and boycotts intensified dramatically. While the Greenpeace campaign increased public pressure on the British government to halt the dumping, it also increased public pressure on the Spanish government to protest against the dumping. Greenpeace was a skillful and resourceful two-level player within the regime.[1]

Though combining a powerful public idea with national and international entrepreneurship was crucial, it should be acknowledged that connections to domestic electoral policies and special circumstances were crucial too.[2] The British Prime Minister, Margaret Thatcher, was already engaged in a high-profile confrontation with the coal miners' trade union and did not want to take on the issue of radwaste disposal issue; Spain's Prime Minister, Felipe Gonzalez, was eager to improve relations with the separatist Basques, who were opposing dumping intensely, and had committed himself to a fierce anti-dumping stance in the London Convention context; and in the United States the Clinton administration's strong support for environmental measures was tipping the balance in favor of the Environmental Protection Agency and the National Oceanic and Atmospheric Administration at the expense of the Department of Defense and the Department of Energy. A complete explanation of the regime change must take into account such domestic and special circumstances in addition to primarily international and global elements.

The changes brought about by Greenpeace set the stage for the sequence of events in which the main actor was the anti-dumping governmental coalition. In contrast to the pre-1972 period, when no regime existed, the dumping regime was an important institutional instrument for the anti-dumping governmental coalition.

The regime change and the decision to ban radwaste disposal were mediated by the global ocean dumping regime in essentially three ways.

First, the regime served as an institutional focal point for governmental and nongovernmental opposition to radwaste disposal. It increased the international visibility of contested government policy. The existence of a permanent global forum in which the issue could be debated from an environmental perspective was clearly advantageous for those who were opposed to radwaste disposal. ENGOs and governments were skillful in using this international institution to protest against radwaste disposal and to present scientific and technical reports in support of halting ocean dumping of radioactive waste. Thus, very much in line with the thinking of UN experts involved in the regime-building process, this environmental regime served as a global forum before which private citizen groups as well as governments could bring their protests concerning dumping practices which were perceived as hazardous and as creating transboundary risks.[3]

Second, within the framework of the regime, the coalition of anti-dumping governments adopted resolutions and later treaty amendments aimed at halting such disposal. As a result of a series of London Convention resolutions that gradually altered the legal substance of the dumping regime in a more pro-environment and precautionary direction, radwaste disposal increasingly became legally controversial. Changes were made with respect to important regulatory issues, including the burden of proof in regard to environmental damage, the underlying regulatory approach, and regulation under conditions of scientific uncertainty. Such a significant legal transformation affecting the regime's principles, norms, and rules required a global regime.

Third, the coalition of anti-dumping governments used the regime to establish global behavioral norms and standards against which individual countries' ocean dumping policies could be compared and judged by other countries, by ENGOs, and by the public. By setting behavioral norms and standards, anti-dumping governments used the regime to increase international political pressure on countries to make their policies environmentally acceptable. Regime norms and standards significantly raised the political costs of noncompliance.

Governments had recognized the importance of peer pressure and public opinion when they equipped the regime with a compliance mechanism in 1972. They agreed that an amendment of the annexes decided by two-thirds of those members present at a consultative meeting would apply to all members except those who made an official declaration rejecting it within 100 days after the decision. In the words of one U.S. negotiator, "it was felt that the procedure adopted would be useful, in that it requires a positive act of refusal, theoretically made more difficult by publicity and peer pressure to accept the proposed amendment" (Leitzell 1973, p. 513). The U.S. delegation had proposed these rules for changes in the annexes (memo, Ministry of Foreign Affairs, Denmark, December 6, 1972, p. 18).[4]

Regime Analysis and the Radwaste Disposal Ban

The three prominent regime approaches propose rather different propositions about how and why regimes change. Realists expect that the global dumping regime would depend on continued hegemonic leadership. In their view, declining U.S. leadership would result in collapse of the regime, because states would follow their own individual interests. But, as described in the previous chapter, despite a lack of support by the United States, at least initially, a significant international policy change with respect to radwaste disposal was achieved. Realists might argue that the lack of U.S. leadership explains why the global ocean dumping regime did not resolutely adopt a precautionary and pro-environmental international policy on radwaste disposal in 1983. It is likely that U.S. support would have influenced the position taken by other pro-dumping governments, and that a sufficiently large number of governments would have accepted a change to the annexes. Moreover, if the United States had determinedly supported the ban, then it might, as in similar international environmental conflicts, have used sanctions of various sorts to pressure pro-dumping governments to follow suit.[5]

In this case, however, the lack of U.S. leadership did not cause noncooperation and regime collapse, and the global termination of radwaste disposal was not a reflection of a change in the underlying distribution of power among states. This case also disconfirms realists' claim that powerful states almost always are able to ignore regimes and can even restructure regimes if they wish to do so. Powerful states lacked necessary domestic support to overrule the global ocean dumping regime or to ignore the radwaste disposal issue. Because realists ignore the domestic level, they cannot predict which foreign environmental policies governments can pursue within environmental regimes.

Neoliberals suggest a benign version of the hegemonic stability theory, according to which declining hegemonic support does not necessarily result in immediate collapse of a regime. Regimes may persist beyond hegemonic support, neoliberals claim, if they provide information, lengthen "the shadow of the future," and facilitate linkage among issues. But, evidently, neoliberals would not expect policy development to happen the way it did in this case. As chapter 8 showed, a transnational coalition used its control over the forum of the regime to establish international rules which large pro-dumping states, owing to the development of domestic policies and protests against radwaste disposal, could not go against. Moreover, it follows from the assumption about anarchy in the international system (i.e., that no central authority exists above states) that monitoring and enforcement of international agreements is largely left to states or to an international secretariat. But this was not the case; an ENGO monitored states.

To judge from this case, neoliberals leave out a good part of what environmental regimes and environmental cooperation is about by assuming that actors are pursuing only their own interests. Unlike those who stress that cooperation fundamentally is directed toward a common goal, they generally downplay the significance of perceived common interests, although they assume that states sometimes might have joint interests.[6] For neoliberals, regimes coordinate interactions among states pursuing self-interest. Robert Keohane's definition of cooperation as the theoretical alternative to "harmony" emphasizes voluntarism and individual gains; states cooperate because it provides them with egoistic gains or rewards.[7] Ideational power, however, can stimulate collective action and thereby provide a common, rather than individual, good.[8] Moreover, and also important, "harmony" does not guarantee that cooperation among nations emerges spontaneously or that influence need not be exercised.[9] Deliberate initiatives aimed at discovering common interests and values and assisting states in identifying joint interests are often necessary in order to realize the potential for cooperation given by a certain set of state objectives and interests.

Land-locked states excluded, states share an interest in radiation-free oceans.[10] But, as was pointed out in chapter 7, realist and neoliberal students of international relations have paid little attention to ideas that can make states define their interests in ways that enhance cooperation. These students would doubt that a change of international public opinion, at least as long as powerful states ignored it, could have a significant impact on a regime and on the way in which states defined their interests. But the radwaste issue had to be constructed before states' interests could be identified. Moreover, it is insufficient to characterize states' interests as purely individualistic and competitive when they agreed to ban radwaste disposal.

Why did scientific knowledge play an insignificant role in this regime change? First, scientific consensus did not support a global radwaste disposal ban; second, there was no significant change in the science in regard to dumping of low-level nuclear waste during the change. For these two reasons, the epistemic-community approach is not able to explain the case adequately.

It should be noted that public opinion and ENGOs were on approximately the same side of the radwaste disposal issue, whereas scientific experts held a very different view. The epistemic-community approach largely ignores ENGOs and the public-opinion variable; this is unfortunate, since public opinion, ENGOs, and scientists all supported stringent environmental protection in the cases examined by epistemic-community analysts: Mediterranean pollution, ozone layer depletion, and protection of whales.[11]

Yet the approach does not separate, measure, and compare those variables, and it is accordingly quite likely that the impact of epistemic communities on international environmental policy has been imprecisely assessed. This case documents the existence of complex relationships of ENGOs, scientific communities, and governments. These relationships should be examined more carefully.

Obviously, the development of the regime did not reflect scientific opinion, and ocean scientists criticized it in professional journals. Pointing out that the 1985 London Convention resolution "totally disregarded" science, the British geobiochemist E. I. Hamilton wrote (1986, pp. 296–297): "There is no scientific evidence to indicate that the discharge of low level radioactive wastes to the sea, land or air is harmful to man." A peer-reviewed 1986 article concluded that the risks from past radioactive ocean dumping in the Northeast Atlantic were "very low indeed" (Camplin and Hill 1986, p. 250). Thus, making a barely concealed reference to the moratorium on radwaste disposal, Camplin and Hill wrote: "It is clear that there are no scientific or technical grounds for excluding sea dumping from consideration alongside other viable disposal options for radioactive wastes" (ibid., p. 251).

Internationally acclaimed scientists who served as advisors to the regime were also critical. The regime's scientific advisory group, GESAMP (Group of Experts on the Scientific Aspects of Marine Pollution, established under UN auspices in 1969) seriously questioned the wisdom of changing the regime's radwaste disposal policy. Members of GESAMP emphasized that, despite the fact that existing knowledge was imperfect and uncertain, the consensus of the marine scientific community was that the risk from past dumping was "exceedingly small" (Bewers and Garrett 1987, p. 118). More generally, GESAMP described the policy development as an example of "lack of confidence in the regulatory process" when full environmental implications of waste emissions were not known (GESAMP 1991, p. 10). GESAMP advised against banning radwaste disposal, but the transnational anti-dumping coalition ignored the regime's scientific advisors, and the scientists had very little political clout.[12]

Specialized UN agencies did not recommend the regime change either. The IAEA compared risks relevant to ocean dumping of low-level radioactive waste and reached essentially the same conclusion as GESAMP. An

160   Chapter 9

IAEA study requested by IGPRAD in 1987 summarized the relative risks to a critical group that, owing to the large quantities of seafood consumed and the higher concentration of contaminants assumed in such foodstuffs, would be exposed to higher doses than any other group of individuals in the population (IAEA 1993, p. 35). (See figure 9.1.) The study showed that the presence of naturally occurring radionuclides could result in a risk greater than that corresponding to the International Commission on Radiological Protection's dose limit for members of the public. Compared to that, the study showed, the additional risks associated with the contamination of seafood from radwaste disposal were about five orders of magnitude lower. Moreover, it appeared that the presence of PCBs could represent a risk similar to that arising from naturally occurring radionuclides. The study concluded that the other chemicals represented risks that were lower, relative

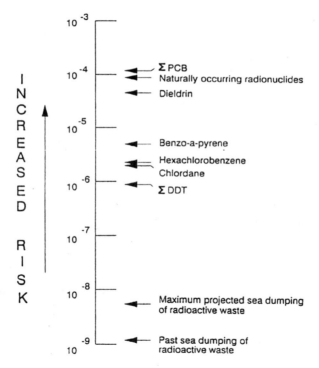

Figure 9.1
Annual risk of fatal cancer induction associated with the ingestion of radionuclides and organic chemical carcinogens by members of the most exposed group. Source: IAEA 1993, p. 35.

to that of PCBs, but still much larger (by more than two orders of magnitude) than those arising from nuclear ocean dumping.

Because of the political and social aspects, the regime's scientific experts were reluctant to become involved in the radwaste disposal issue. GESAMP declined IGPRAD's request to develop operational definitions of such terms as "harm," "safety," "proof," and "significance" to be used in studies and assessments called for in the 1985 resolution on the ground that such definitions, in addition to scientific aspects, involved nonscientific aspects outside the terms of reference of GESAMP (LC/IGPRAD 1993, Annex 2, p. 36). Furthermore, because it based its expert advice on the concept of assimilative capacity, the expert group could not address concerns about radwaste disposal raised by many regime members.[13] Governments ignored GESAMP's expert advice as they wished.

Scientific experts distanced themselves from the highly politicized issue of radwaste disposal. At the same time, governments and ENGOs took steps to reduce the influence of marine scientific advisory groups. As already pointed out, the London Convention originally stipulated that global ocean dumping regulation should be based only on technical or scientific considerations and thus ensure that dumping regulations were based on the advice of marine scientists. At the level of regime principles and norms, however, the 1991 consultative meeting's decision to substitute the concept of assimilative capacity with the precautionary principle significantly reduced marine scientists' role as policy experts within the regime. If the regulatory goal was to reduce waste discharges as much as possible, then regulatory decisions would be concerned with choosing technologies that best met this goal. Marine scientists could not contribute to the realization of this goal (Clark 1989a, p. 295). Probably also for this reason, GESAMP emphasized that its approach to protection of the marine environment assumed the need for caution; however, the expert group did not endorse the precautionary principle, because it found that the principle could not provide a scientific basis for marine pollution control (GESAMP 1991, p. 8). Noting the increasing international prominence of the precautionary principle, one GESAMP scientist even described a general "declining influence of science on marine environmental policy" (Bewers 1995).[14]

Most governments responded to public concerns about ocean dumping of radioactive waste and wanted to protect their fishing, environmental, tourism, and related interests against the risk of radiation, however small.

A British regulator noted: "I think everybody who objects to sea dumping accepts that the risks that are present look as if they are low, and it is not really an argument on that basis at all. It is simply that, for example, people who get their living from the sea do not like anything put into the sea, they see it as a threat to their livelihood, and it becomes much more that kind of argument than anything at all to do with radiation."[15] Since the 1970s, a concern for the health of the marine environment originating in the United States and parts of Western Europe was increasingly shared by developing countries, especially those dependent on the ocean for their livelihood. It created an international trend toward elimination of perceived threats to the health of the ocean.[16]

Experts and scientists might play important roles in international environmental regimes, but it should not be overlooked that governments will not for long champion environmental policies that are justifiable on scientific and technical grounds but unacceptable to the general public. In this case, governments, environmental protection agencies and the public were unwilling to accept small risks with potentially large consequences—so-called low-probability, high-risk issues—and took nonscientific aspects into account as well. Environmental protection agencies found that land disposal appeared to be less risky from the perspective of monitoring and retrieval than ocean disposal. Furthermore, as government agencies, they wished to compel society to find more effective ways of waste handling and waste reduction; as often pointed out by environmentalists, ocean dumping was an "easy out."

Theories of ENGOs' Roles in International Environmental Politics and Regimes

There is today a growing academic literature about NGOs, and ENGOs constitute an emerging field in international environmental politics.[17] Many observations and theories are suggested in this literature. From an environmental protection point of view, some analysts are encouraged by the emergence of ENGOs, which they view as new forms of political organization embodying a transnational ecological consciousness. Others claim that ENGOs' participation in international environmental diplomacy does not signify that states have become less powerful or have less control over outcomes of international environmental politics.

Scholars have paid special attention to the importance of civil society. Paul Wapner (1995), who focuses primarily on ENGOs' involvement in civil societies and markets, emphasizes that their impact on government policies is only a minor facet of their political and ideological impact on global politics. He takes a sociological approach to how ENGOs disseminate an ecological sensibility, pressure corporations, or empower local communities, and thereby change world politics. Because he focuses on extra-state spheres, Wapner makes only implicit claims about state-ENGO relationships; however, he suggests that ENGOs influence states through such activities.

Ronnie Lipschutz also emphasizes the importance of global civil society. He sees Greenpeace as one participant in the networks of global civil society, and regimes as serving "the specific interests of state and governments" (Lipschutz 1992, p. 397). Accordingly, ENGOs would play an insignificant role within the global dumping regime, while states would use the regime to realize their interests. As chapter 8 demonstrated, however, these propositions are not supported in the case of the radwaste disposal ban.

Kal Raustiala, who focuses on ENGOs' participation in international environmental diplomacy, claims that ENGOs do not supersede states in international environmental politics. His explanation of how states benefit from ENGOs' participation in environmental negotiations and regimes combines functionalism, neoinstitutionalism, and rationalism. His main assertions—that "the specific forms of NGO participation granted are systematically linked to the specialized resources NGOs possess" and that states gain advantages from NGOs' participation—create a challenge for other studies concerned with ENGOs (Raustiala 1997, p. 734). Raustiala identifies six ways in which ENGOs might assist governments involved in environmental negotiations and international institutions: policy research, monitoring of state commitments, "fire alarms," negotiations reporting, revealing the "win-set," and facilitating ratification. He would expect that states benefited from and controlled ENGOs' participation when the global ocean dumping regime changed to emphasize precaution and prevention in regard to radwaste disposal. But again, these propositions about ENGOs' participation in international environmental diplomacy are not confirmed in the case of the global radwaste ban.

In their co-edited volume *Environmental NGOs in World Politics*, Thomas Princen and Matthias Finger comprehensively critique realism and power-based theory of international relations. They theorize about the "NGO phenomenon," and they present much empirical evidence of the worldwide growth and diversity of NGOs. They see NGOs as agents of change who play a critical role in social learning and in connecting world politics and biophysical changes.

Princen (1994, p. 32) assumes that regime-change processes would not be led by states. According to Princen and Finger (1994), the technical nature of the issue and the analytical processes needed to protect the environment also reduce the role of diplomats in environmental diplomacy. Princen predicts that the main players in regime development are epistemic communities, individual leaders, and ENGOs, but does not identify more precisely the roles these actors will perform. He assumes a convergence of interests among epistemic communities, prominent individuals, and ENGOs, and he would expect them to influence the global ocean dumping regime to put more emphasis on the environment; states would be insignificant bystanders.[18] With regard to the actions of Greenpeace International, Princen (ibid., p. 33) claims that the organization "identifies a problem area, enters for a direct action protest, gets the media coverage, and then disappears." Thus, in his opinion, it is not likely that Greenpeace International would be involved in the global dumping regime for a longer period, or that it would organize local constituencies. But the global radwaste disposal ban shows that Greenpeace did exactly what Princen expects it not to do: the organization was involved in the issue on a permanent basis, and it established a transnational environmental coalition that included local, national, and regional stakeholders and interest groups.

Toward a Theory of ENGOs' Influences on Regimes

Chapter 8 showed that an ENGO acted as a catalyst for regime change within the global dumping regime and forced the change from control to prohibition. Greenpeace, a well-staffed, professional, global ENGO, mobilized national and international public opinion, and it strengthened a transnational coalition by mobilizing other ENGOs, stakeholders, special-interest groups, and mass constituencies. Moreover, Greenpeace attacked scientific and regulatory principles and norms of the regime, and it monitored compliance by states. Four hypotheses about the roles of ENGOs in regime development can be generalized on the basis of this case.

First, *ENGOs influence regime development by mobilizing public opinion*. Public concern makes it politically costly for governments to ignore perceived environmental-protection needs. It can also result in tougher environmental-protection measures. Public opinion raises the political costs of disregarding international environmental regimes; put emphatically, "governments that are reluctant to sign will be pilloried at home."[18]

Mass publicity has been critical for the ENGOs established in the late 1960s and the early 1970s (Hansen 1993, pp. 150–178). Successful public diplomacy depends on mass media effectively communicating powerful images and ideas to policy makers and the general public. "Pollution ideas are," as Mary Douglas and Aaron Wildavsky have pointed out (1983, p. 47), "an instrument of control," and the essence of effective campaigning is to use powerful metaphors, images and ideas with great effect. "Save the whales," "the oceans are dying," and "the ozone hole" are examples of influential images and ideas that have been backed by environmentalists and have played a significant intellectual role in global environmental politics.

The availability of powerful public ideas and the deliberate construction of metaphors and symbols of serious environmental damage and crises is essential in public diplomacy. Therefore, in their absence, it is unlikely that ENGOs can sell or market an issue to the public, and there will be little room for persuasion, campaigning, and public diplomacy by ENGOs. Following a similar line of reasoning, scholars who emphasize crises and exogenous shocks expect little progress on issues such as international management of biodiversity and global warming, at least as long as only gradual environmental damage is occurring.[20] A metaphor highlighting demonstrable damage due to global warming (e.g. a "climate hole") is most likely necessary in order to focus sufficient attention on global warming and persuade governments to act aggressively against this problem (Schneider 1998, p. 18).

Greenpeace's public diplomacy has often been spectacularly effective. By staging spectacular "happenings" or series of focusing events (so-called campaigns), Greenpeace has focused the attention of the international public and mass media on ocean dumping of radioactive waste, nuclear testing, whaling, sealing, oil rig dumping, and other "environmental crimes." "We want to draw attention to something," a founder and chairman of Greenpeace explained (Thomas 1984). "We use action and, once there's attention, we move into lobbying." Greenpeace is probably the main "green" influence on international public opinion, but other ENGOs also shape public opinion.

Second, *ENGOs influence regime development by transnational coalition building.* Strong transnational coalitions are built by extending ENGO networks and by adding scientists, special-interest groups, and governments. Special-interest groups affected by environmental deterioration dramatically alter the configuration of interests and significantly strengthen transnational environmental coalitions. It is highly likely that special-interest groups, as opposed to the general public, will be part of a transnational environmental coalition, because those whose health or income are directly affected by environmental deterioration are strongly motivated to act. This is confirmed in many environmental conflicts, including game hunters pressing for protection of elephants and the tourism industry's protesting against pollution of the Mediterranean Sea.[21] Furthermore, it is of great significance in environmental diplomacy that the number of states in a transnational coalition group be large, as this makes it possible to pass resolutions, moratoriums, and global bans. ENGOs play a significant role in recruiting new members, especially downwind and downstream states, in transnational environmental coalitions.

Transnational coalition building, which figures prominently in the case of the radwaste ban, deserves much more attention. As discussed above, some of the recent literature draws attention to the importance of civil society. Though the case of radwaste disposal could be seen as confirming the claim about the political significance of global civil society made by some scholars, it also shows that ENGOs significantly increase their influence on regime development by directly "bringing in" interest groups and civil society actors. Knowledge-based regime analysis pays much attention to transnational networks of scientists, but coalition building among influential special-interest groups (in which ENGOs play a major role) is not yet well understood. Shipbuilders, classification societies, and insurers have been important in creating an effective compliance system for controlling intentional pollution by oil tankers, and in the case of global warming some ENGOs see the insurance industry as an influential interest group to be included in a coalition of interests that can effectively challenge the fossil fuel industry (Mitchell 1994).[22] However, coalitions are not necessarily enduring or stable. At an early stage of international regulation of chlorofluorocarbons (which deplete the atmospheric ozone layer), the chemical company DuPont opposed regulating CFCs as this would hurt the company economically, but DuPont later urged regulation as this would serve its long-term economic interests.[23] Generally, different distributions of losers and winners create different opportunities for building and sustaining national and transnational coalitions.

Third, *ENGOs influence regime development by monitoring environmental commitments of states*. For example, in the case of the endangered species trade regime (CITES), the World Wide Fund for Nature (WWF) has conducted monitoring and some degree of enforcement (Princen et al. 1994, pp. 217–236).[24] ENGOs are an important part of environmental governance on an international scale, and as possessors of local knowledge they provide governments with useful information. One Norwegian politician said the following about Bellona (an ENGO monitoring the nuclear waste situation in the northwestern part of Russia): "Bellona is a supplement and has shown that a voluntary organization enters environments and gets information not available to a public authority. For this reason, I believe in continuing the cooperation between private and public organizations." (Mathismoen 1994; translation by L. Ringius) ENGOs also generate information on compliance and noncompliance, which they use to great effect, and ENGO monitoring may compel states to revise their policies. For example, ENGO monitoring in the Antarctic was followed by protests that pressured the French government to shelve its plans to build an airstrip (Laws 1990, pp. 121–149).

The radwaste disposal ban demonstrates that monitoring by ENGOs is significant and that ENGOs generate information on compliance and noncompliance, which they use to great effect in global environmental politics. As discussed above, other studies also ascribe importance to ENGO monitoring. The issue deserves to be examined more fully, and it could become apparent that this source of influence depends on combinations of the specific nature of individual environmental issues and ENGOs. ENGOs' monitoring capability should be expected to be highly dependent on their organizational resources. Also, the scope of environmental problems may facilitate or reduce monitoring by ENGOs. For example, as early as the 1920s experts considered it very difficult to detect oil slicks and to link them with the responsible tanker (Mitchell 1994, p. 446). It is also evident that obstructing and boycotting states' environmental policies is not always possible. For example, Norwegians generally support controlled commercial whaling, and the Norwegian state has resolutely protected Norwegian whalers against ENGOs' interference within territorial waters ("Greenpeace must pay," *Aftenposten*, March 15, 1995; translation by L. Ringius).

Fourth, *ENGOs influence regime development by advocating precautionary action and protection of the environment*. ENGOs that are looked upon as channels and legitimate representatives of public opinion and environmental interests have often participated in meetings within environmental regimes. Increased ENGO presence on national delegations and ENGO "delegations" at international meetings have shifted the burden of proof onto polluters, and ENGOs' participation has resulted in more emphasis being placed on precaution and nonscientific and ethical factors.[25] ENGOs often participate in scientific and technical working groups established by regimes, which in turn offers them additional opportunity to emphasize their concern for the environment, even in the absence of conclusive scientific proof of environmental damage.

ENGOs are highly influential when the often vague norms and principles of environmental regimes must be operationalized in order to guide the formulation of more precise regime policies and policy goals. Interpretation and operationalization will unavoidably involve value judgments and subjective forms of risk assessment (Majone 1989). Governments may be under pressure to allow ENGO participation, but may also find ENGO participation desirable because ENGOs counterbalance national and international special-interest groups and because their participation may serve to legitimize regime rules and regulations. Moreover, there is an unmistakable tendency toward creating environmental-protection rules and norms at the level of the lowest common denominator. However, by strongly advocating precaution, ENGOs expand transnational environmental coalitions; this attracts governments (e.g., those of Kiribati and Nauru) that join only when environmental protection is given high priority by regimes.

The global radwaste ban illustrates the significance of ENGOs' advocacy of a precautionary approach to protection of the environment. Scholars often overlook this form of influence, perhaps because of its more discursive nature, or because its significance might be more difficult to document accurately. But sometimes ENGOs supply concepts and play important roles in framing international environmental issues. In the case of the Antarctic, Greenpeace backed the idea of a World Park (Princen et al. 1995, pp. 52–53); this illustrates that ENGOs can influence regime development by producing ideas and systems of thought that become accepted as parts of the intellectual foundation of environmental regimes. Again, it might become apparent that this form of ENGO influence varies in a systematic fashion from one environmental issue area to another and among ENGOs.

The four hypotheses about the roles of ENGOs do not presume that ENGOs are powerful and influential in global environmental politics while states are powerless and without influence, or the reverse; instead they shed light on subtler but nonetheless important relationships among ENGOs, states, and epistemic communities. It should be expected that the ENGOs that perform all four roles are the most influential ones. Moreover, an ENGO's capacity to influence regime development depends on its organizational resources (staff, finances, expertise, etc.) and on its strategy.[26] The four hypotheses differ significantly from those of prominent approaches to regime analysis, and to some extent also from recent studies on ENGOs and international environmental politics.

Conclusions

I have shown that the case of the global radwaste disposal ban contradicts claims made about regime change in power-based, interest-based, and knowledge-based regime analyses. I have, in addition, shown that the empirical findings do not confirm propositions suggested by prominent recent studies of ENGOs and international environmental politics. To gain a better grasp of the sources of ENGOs' influence on international environmental regimes, I have suggested four hypotheses as to how ENGOs influence regime development by mobilizing international public opinion, building transnational environmental coalitions, monitoring compliance, and advocating a precautionary approach to environmental protection.

The case of radwaste disposal raises some fundamental questions about the roles of ENGOs and, more generally, whether some environmental issue areas and regimes are particularly susceptible to the involvement of ENGOs. The four hypotheses about ENGOs' roles in regime development are supported by this case and by other cases. The significance of ENGOs' monitoring of government policies and regimes is noted in some of the analytical literature on ENGOs, but it is not fully acknowledged that ENGOs influence regime development by mobilizing international public opinion, building transnational coalitions, and advocating preventive actions and environmental protection.

Recent studies have focused attention on ENGOs, but regime change and interaction between ENGOs, states, and epistemic communities will require further examination. Clearly, single-actor explanations of regime development will not be sufficient. It will also be useful for analysts to think about different stages of regime change and, more broadly, about different processes of regime change. Comparative empirical research on international environmental regimes will be useful to examine the four hypotheses about ENGOs' roles in environmental regimes and international environmental politics.

10
Conclusion

Since the 1960s a number of regimes have been created in response to environmental concerns, particularly those that have emerged in industrialized countries. These regimes are evidence of an increasing internationalization, even a globalization, of environmental policy and regulation. At present it seems highly unlikely that a world government with policy force and judicial power will develop in the foreseeable future, and future environmental regimes will probably come about in ways that are not significantly different from the ways in which regimes were created in the past.[1] It also seems a safe proposition that countries will continue to use regimes to attempt to solve transboundary and global environmental problems. Although some would like them to be more powerful instruments of environmental protection, regimes do under some circumstances stimulate the development of international and global environmental policy by creating joint gains and, at the same time, constraining governments' freedom in the environmental field.

In chapter 1 I pointed out that norms and principles establishing the deep normative structure of regimes are considered important in regime analysis. But relatively little has been established about the processes through which regime norms and principles are developed, especially their intellectual and substantive content. The same is true with regard to the relationships between widely accepted ideas and global environmental regime norms and principles. Public ideas and actors and channels expressing as well as molding public ideas should play a more prominent role in environmental regime analysis. At the very least, this study cautions one to be skeptical about environmental regime studies emphasizing power, narrow egoistic self-interest, and scientific knowledge at the expense of everything else.

The Global Ocean Dumping Regime, Regime Analysis, and Transnational Coalitions of Policy Entrepreneurs

A change in environmental ideas, values, and norms, not a change in economic and political interests or in scientific knowledge, essentially explains why the global ocean dumping regime was established. American ecologists and environmentalists had been warning about marine pollution problems at least since the mid 1960s, but it took a series of dramatic incidents before legislators began to establish domestic ocean dumping regulation. Ocean dumping, a visible but relatively minor source of marine pollution, was perceived to be a rather urgent and significant environmental problem. In addition, it was conceived of as a truly international, if not global, environmental problem, especially in the United States. Industrialized countries were dumping unknown amounts of waste, which ocean currents apparently mixed and transported to shores far away from where they had originally entered the ocean. The common understanding was that many countries contributed to this environmental problem and that all countries would feel the negative effects. Despite scant scientific evidence of damage to the marine environment from ocean dumping, a transnational coalition of policy entrepreneurs whose core consisted of prominent environmentalists and scientists, U.S. legislators, and the Secretary-General of the Stockholm conference, together with his secretariat, convinced and persuaded countries that a global ocean dumping regime would be in the interest of all countries. By spreading a powerful idea ("the oceans are dying") that changed hegemonic and international perception of ocean dumping of waste, this transnational entrepreneur coalition initiated and structured regime formation. It also developed regime features that solved important scientific and distributive issues jeopardizing achievement of agreement on the regime. The hegemon, where the transnational coalition had its origins and where it was most politically influential, proposed to other countries that a global regime to control ocean dumping be built.

Let me recapitulate the shortcomings of power-based regime analysis: This case showed no evidence supporting that power in the material sense was of fundamental importance. The United States did not make use of its material and political superiority to pressure other countries to join the global ocean dumping regime. Instead, the United States set a good example for other to follow. Regarding the interests and motivations of the hegemon, moreover, the United States perceived that its own interests and other countries' interests were deeply intertwined on this global environmental problem. Legislators who were convinced that the United States had contributed considerably to the ocean dumping problem supported U.S. ideational leadership in the regime-building phase. Notably, legislators even felt some measure of guilt for past dumping by the United States. Evidently, such findings conflicted with the realist claim that individuals and states are best understood as egoists. The case raised important questions about the role and the significance of self-interest, altruism, and common interests. Other scholars find it at least equally true that "in actual situations people often do not follow the selfish strategy" (Sen 1977, p. 341)—people sometimes behave in accord with social norms, show concern for others, and even might act altruistically.[2]

The interest-based regime model was much better able to account for the establishment of the global ocean dumping regime, especially by putting emphasis on integrative bargaining and the significance of entrepreneurial leadership provided by Maurice Strong and the Stockholm secretariat. Strong and his secretariat were instrumental in identifying global marine pollution as an area of common concern and ocean dumping as an area in which a global environmental regime was politically feasible. The black and gray lists developed by UN experts and promoted by the secretariat accelerated global environmental negotiations, solved the problem of insufficient scientific knowledge, separated science from politics, and, further, represented a piecemeal, gradual approach to ocean dumping regulation. Essentially, the system of black and gray lists accommodated the interests of pro-environment industrialized countries and those of less-concerned developing countries.

It was also evident, however, that neoliberal analysts focused too exclusively on negotiations among government representatives. Significant ideational processes were initiated in parallel with the negotiations on the global dumping regime. Strong, the Stockholm secretariat, and ecologists and prominent scientists communicated the seriousness of marine pollution caused by ocean dumping to national negotiators and mass audiences. Strong and his secretariat played a critical role in the political construction of the ocean dumping problem at the global level. Yet interest-based propositions about individual intellectual leadership take a more narrow approach to negotiation and to the behavior and perceptions of government negotiations. Neoliberal propositions pay little attention to communicative forms of leadership, to public diplomacy, or to the mobilization of world public opinion. Bargaining is of course important, but it takes place only when the issue has been constructed.

Despite useful insights in the regime literature concerned with scientific and technical knowledge, its emphasis on consensual knowledge created too narrow a view of the potential political influence of environmentalists and scientists. This case showed that backing by a sympathetic public opinion was at least as significant for influential policy entrepreneurship by prominent environmentalists and scientists as whether they could conclusively prove that a serious environmental problem really existed. In spite of scientific uncertainty and dispute, distinguished environmentalists and scientists initiated and framed political debate and motivated policy makers to launch new policy initiatives. But knowledge-based analysis ascribed little importance to ideas-based actors others than epistemic communities and knowledge brokers. Furthermore, as political legitimacy and public support for policy were ignored, little attention was paid to public opinion. In view of the emphasis put on ideas and perception in knowledge-based analysis, these are significant shortcomings. It is unfortunate, though not entirely surprising, that reflectivists and epistemic-community theorists largely overlooked how persuasion and communicative action influenced regime processes in this case. This study showed that attracting the attention of international mass media and, in turn, mobilizing and shaping national and international public opinion was a crucial element of the strategy to influence global environmental policy pursued by U.S. legislators, the Stockholm secretariat, and a global ENGO.

In summary, since they overlook some of the significant actors and processes, power-based, interest-based, and knowledge-based regime analyses could neither singularly nor in some combination account adequately for regime formation in this case. These three approaches exhibit some significant similarities and common weaknesses. They pay little attention to widely accepted ideas and policy entrepreneurship, and the opinion of the public—in other words, the attitude of the ultimate stakeholder—is generally ignored in mainstream regime approaches. Only few contributors to regime analysis give much weight to the creation of legitimacy, to the justification of policy, or to the mobilization of public support. The approach concerned with transnational coalitions of policy entrepreneurs, in contrast, expanded regime analysis to include the broader political arena, state-society relationships, and the significance of state-society relationships for policy and regime development. The strength of the TEC approach is that it connects three important factors in environmental cooperation: policy entrepreneurs, ideas, and the public arena. The TEC approach combines a focus on an influential transnational actor with ideational analysis of regime processes.

I have also demonstrated that ENGOs play a crucial role in changing international environmental norms, principles, rules, and decision-making procedures, even in the face of opposition from dominant states and lack of scientific proof of environmental and human health risks. Since the mid 1980s, the global regime on ocean dumping of low-level radioactive waste has undergone a major change from a permissive allowance of such disposal to a total ban on disposal that reflects a recent emphasis on precaution and prevention. A global ENGO mobilized national and international public opinion, broadened a transnational environmental coalition, monitored environmental commitments of states, and advocated precautionary action and protection of the environment. It was evident that theories emphasizing either material power, self-interests, or knowledge could not account adequately for the processes and actors that accomplished this regime change. In particular, they paid little attention to the underlying ideational and normative dynamics of this regime.

Ideas, Hegemons, and Regimes

In this book I have examined how transnational coalitions of policy entrepreneurs and public ideas interact in processes constructing or reconstructing state interests, international environmental problems and issue areas. I have paid less attention to what, if any, impact such transnational ideas-based actors might have on relations among states. Reflective and ideational scholars often seem to believe that the existence of ideas-based actors is evidence of a diffusion of sovereign power of states to transnational actors and, by implication, that significant consequences for world order follow.

Do environmental regimes built by transnational entrepreneur coalitions benefit some states more than others? Do these regimes distribute costs and benefits differently than regimes that come about through processes in which power, interests, or knowledge play the predominant role? We do not yet know the answers to those questions, and it might not be possible to answer in general. It is nonetheless useful to briefly examine the creation of the global ocean dumping regime in the light of the interests of the United States (the hegemon). As was discussed in chapter 3, it should be recalled that ideas and ideas-based actors should be expected to play a role in situations where states are dealing with efficiency issues and are trying to reap joint coordination gains, but not in distributive conflicts among states.

As has been pointed out repeatedly, a concern for marine environment protection was the driving force behind establishing the global ocean dumping regime. But broader American economic and political interests also supported establishing a global regime in this issue area. Not insignificantly, the United States demonstrated that it took the environmental question seriously in Stockholm by proposing concrete action on the ocean dumping question. And, to recapitulate the nonenvironmental benefits to the hegemon, global environmental standards that could be agreed to under a future regime would harmonize pollution control costs associated with ocean dumping. The establishment of the global ocean dumping regime at a point where domestic regulation threatened to disadvantage American industry therefore meant that the United States succeeded in protecting its economic interests. By establishing a global regime, the United States also satisfied another interest unrelated to the specific issue of ocean dumping, namely its political interests in designing institutions for countries' interaction. American politicians were pivotal in constructing a view of ocean dumping that would back the environmental policy they were creating and would at the same time protect national economic and political interests. This indicates that well-recognized state interests establish the basic parameters determining the scope for cooperation but that ideas can facilitate cooperation as long as it develops within those parameters.

As I have documented, the establishment of this environmental regime unfolded almost entirely in parallel with the establishment of U.S. domestic ocean dumping regulation. It is simply not possible to understand why this regime was created without including changes in U.S domestic regulation. In a situation where an idea was compatible with its political and economic interests, the United States once again demonstrated that it is the "state most inclined and most able to project its domestic political and economic arrangements onto the world" (Burley 1993, p. 125). When examined at the level of states, rather than the level of transnational actors, this case shows that the hegemon achieved a regime serving its combined environmental, economic, and political interests. In the final analysis, therefore, this regime grew out of a self-initiated regulatory change in the United States. At the intergovernmental level, the case therefore could suggest a hegemonic explanation for the very existence of this global environmental regime.

Yet the hegemon created the regime neither through military force nor by offering rewards or side payments; instead, it projected the politically constructed rationale for domestic regulation onto the international level. Similar to the domestic-level process in the United States, policy entrepreneurs pushed a powerful public idea (the "dying oceans" idea) identifying a seemingly evident and indisputable need for international environmental regulation. U.S. leadership was of an ideational nature, and it seems best captured by a concept of hegemony stressing material and structural resources as well as cultural norms. The Gramscian concept of hegemony in particular emphasizes the hegemonic state's ability to maintain cohesion and identity within a core group of states through the propagation of a common culture.[3] In addition, this concept of hegemony envisions that international organizations, such as the United Nations, assist in spreading and consolidating values and norms of the hegemon. Accordingly, the Gramscian concept of hegemony can also account well for the important role the Stockholm secretariat played during the process of regime establishment.

Though it is neither possible nor advisable to draw broad generalizations on the basis of a single-case study, it is not too difficult to point to other examples of such U.S. leadership and hegemony in a nonmaterial, ideational sense in the environmental field. For example, a former U.S. ambassador to the United Nations had such ideational hegemony in mind when he observed, as mentioned in chapter 6, that the U.S. government and American elite groups in the early 1970s had become great producers and distributors of world environmental crises. Examples also exist in other policy fields. In the area of macroeconomic policy, for example, the United States spread Keynes's economic ideas to other countries after World War II. Stressing American military and economic dominance at that point, Hirschman generalizes this hegemon-initiated ideas process as follows

(1989, p. 351): "It would seem that, to achieve worldwide influence, an economic idea must first win over the elite in a single country; that this country must exert or subsequently chance to acquire a measure of world leadership; and that the country's elites be motivated and find an opportunity to spread the new economic message." Certainly, Keynes's economic ideas and the public idea of "the dying oceans" are different sorts of ideas and theories. Nonetheless, the intellectual hegemony of the United States seems strikingly similar in those two instances of international dissemination of ideas.

I have examined ideas as a means of persuasion aimed at increasing global environmental protection by constructing global environmental norms, principles and institutions for policy coordination. But it is by no means evident that realism, except in its most orthodox formulations, necessarily must ignore the power of ideas. Quite the contrary, it is a reasonable assumption that political leaders are quite aware that ideas might play an important role in furthering national interests. Senator Warren Magnuson, for example, underscored the political significance of ideas in 1970 when he submitted a resolution to the Senate to create a world environmental institute: "Surely the time has come for the United States to take the lead and propose creation of the Institute to the nations of the world. The time has come for us to realize that world leadership and world prestige are based on the power of ideas, not on the power of weapons." (Magnuson 1971, p. 131) Ideas are under some conditions an instrument of control and might be used instrumentally, for example to realize national interests.

A final observation on the relationships between power in the material sense, ideas and international norms is in order. As this study demonstrated, a change in U.S. policy was almost instantaneously paralleled by a change in global policy. This case therefore indicates that states are not equally able to define the scope and purpose of global environmental regimes. Global environmental problems causing concern to a hegemon—or to a group of powerful states—are likely to be at the top of the international environmental agenda. A hegemon—or a group of powerful states—often assisted by an international organization, is much more able to put its imprint on global institutions for environmental protection than are less powerful states. The environmental ideas and norms of dominant societies might not just be internationalized but also to a large extent internalized by other societies throughout the world.[4] Thus, environmental ideas and norms float from the powerful states to the less powerful states, but only rarely in the reverse direction.

Limitations of This Case Study

It is always necessary to be careful when drawing generalizations from a single-case study. First, the global ocean dumping regime was built in a period of heightened concern for the environment. In the late 1960s and the early 1970s, the United States was in the midst of an "environmental revolution." For the first time, a number of environmental initiatives were launched in the United States and other industrialized countries. Institution building for environmental protection at the international level had previously been modest. In such circumstances, it should be expected that environmental ideas could have a larger-than-normal impact upon societies and governments, and that some of the impact of the "dying oceans" idea is explained by the environmental revolution then taking place.

Although this seems intuitively correct, an upsurge in broader concerns for the environment does not explain why the particular issue of ocean dumping became a high-priority environmental policy issue. Moreover, every accident and mishap does not result in calls for government intervention or in legislative action. Conversely, it could be argued that the concern for the health of the oceans was a prominent environmental issue in this period. It nonetheless seems likely, because of the novelty of the environment and ecology at the time, that a feeling of urgency and crisis meant that the "dying oceans" idea made a greater imprint than would otherwise have been the case, and that this accelerated and facilitated the establishment of a regime.

As I have described, the United States took steps to deal with ocean dumping sooner than most other countries. Instead of interpreting this in terms of leadership, it could be seen as an indicator of the greater susceptibility of the American political system to environmental ideas as well as its ability to quickly enact environmental policy. Comparative studies have documented that the United States has reacted sooner, but not necessarily more effectively, than other countries against perceived environmental threats.[5] Unilateral U.S. leadership could therefore be understood as a reflection of the greater susceptibility of the American political system to environmental ideas and concerns. Needles to say, the quicker response of the United States to the ocean dumping issue was a precondition for the American leadership style based on powerful ideas, communication, persuasion, and setting an example for others to follow. But the shorter "response time" of the American political system tells us nothing about why the United States chose to act as a leader in this case.

As I have documented, Maurice Strong (the UN Secretary-General of the Stockholm conference) and the Stockholm secretariat were closely involved in a number of ways in the negotiations on the global dumping regime, making a significant ideational impact as well. As regime formation unfolded in parallel to the Stockholm conference, there existed a good opportunity—an open policy window—for high-level international officials and an international secretariat to get involved in the political construction of the ocean dumping issue and in international public opinion formation. Various spillover effects from the Stockholm conference to the regime establishment process occurred and, among other things, the pressure on negotiators to reach an agreement on a global dumping regime increased as a consequence of these. Counterfactually speaking, there would have been less pressure on negotiators if the regime-building process had not been closely intertwined with the Stockholm conference. But these spillovers were to a significant degree generated by Strong and his secretariat. Thus, Strong and his secretariat influenced both the broader political and ideological context as well as regime formation viewed more narrowly.

More generally, the empirical findings in regard to the regime-building process show that a significant portion of the empirical research focused on the domestic policy process of the hegemon. This research choice is justified by the important role ascribed to the hegemon in power-based regime analysis and international relations studies more generally. To understand the foreign policy of the hegemon, it is necessary to carefully examine the domestic policy process. Nonetheless, the theoretical argument presented about policy entrepreneurs and public ideas might be somewhat more relevant for understanding the domestic policy process in the United States than in other countries. As discussed below, it would therefore be useful to examine systematically the way in which access points and opportunities for coalition building for domestic, transnational, and supranational entrepreneurs vary across national political structures.

It should also be acknowledged, although attention was paid to the pollution control or abatement costs of ocean dumping regulation, that the costs of environmental regulation did not attract as much attention in the early 1970s as they did later. Other studies confirm that relatively little attention was paid to costs of environmental protection in the early 1970s.[6] Governments and the private sector have learned since then that environmental regulation, which in an early stage might appear innocuous, may later turn out to be costly. On the one hand, this could make states less likely to cooperate if they believed that cooperation was unacceptably costly and that the environmental-protection benefits did not justify the pollution-control costs. On the other hand, however, harmonizing the cost of environmental policy and environmental regulation across countries is often an important driving force behind international environmental policy coordination.[7]

As was explained in chapter 1, this study focused on radwaste disposal because the emerging controversy around this issue made it a significant and fascinating case study. How we choose to dispose of our radioactive waste is an issue of paramount environmental significance. Furthermore, the radwaste disposal issue was of great significance for the functioning of the regime, a fact which also made it an appealing research topic. In addition to low-level radioactive waste, the global ocean dumping regime deals with a number of other waste forms and disposal forms (e.g., incineration at sea). Though the analysis of the regime-formation process covered all substances regulated under this regime, it could be seen as a limitation that the study of the regime transformation process focused exclusively on radwaste disposal. Although this is to some extent true, this limitation could be remedied by comparing the results of this study with studies examining other substances regulated under this regime.

The public, political leaders, and ENGOs perceive radioactive waste and other aspects of nuclear technology mostly negatively.[8] What does this mean for the generalizability of the findings? It means that some caution is called for. It should be expected that policy entrepreneurs could quite easily persuade the public that ocean disposal of radwaste is a horrid anti-environmental activity. Because international public opinion is likely to perceive radwaste disposal negatively, mobilizing domestic and international antidumping sentiments would be reasonably straightforward for an influential global ENGO. But it would be very hard to convince the general public that sea disposal is an environmentally neutral and safe, and perhaps even an environmentally preferable, disposal option for this radioactive waste.[9]

It might be argued that the circumscribed role of the regime's scientific expertise is a reflection of the public's fear of radioactivity. Examination of other international environmental issues might conclude that international public opinion sometimes is a weak source of influence. Public opinion, national and international, may also be divided on and even inattentive to particular issues.[10] Nevertheless, as I have suggested in this book, whether this is the case or not will to a significant degree depend on transnational policy entrepreneurship and the availability of ideas, images, and metaphors able to attract public and political attention.

It should be pointed out, finally, that the radwaste disposal issue polarized countries into basically two groups: a large anti-dumping group and a much smaller but materially powerful pro-dumping group. This issue therefore created a particular dynamic that could explain the political attractiveness of moratoriums and bans. This is not unimportant in understanding the dynamics of this issue, but it cannot fully explain the international policy development in regard to radwaste disposal.

Recommendations for Research

I have not claimed in this book that power in the material sense and interests are insignificant in regime development, nor that only ideational factors matter. From an ideational point of view, interests and power clearly matter; however, they cannot alone explain the content of specific policies. It is insufficient to show that a policy satisfies interests; policies have also a cognitive content. At the same time, ideas and theories alone are not powerful enough themselves to determine policy development, and ideational analysis should therefore not be seen as an alternative to approaches emphasizing power and interests. The analytical ambition of regime theory should be to integrate ideas and interests, rather than to segregate them. By focusing on the role of ideas, it is possible to develop auxiliary hypotheses that supplement well established theories about how power and interests influence regime development.

It is necessary to carefully examine how, when, and why ideas matter in environmental cooperation and environmental regime development. I have sought to identify with more precision the conditions under which ideas are likely to be influential. I have also suggested a number of propositions about why one set of ideas had more force than another in a given situation, and in what way particular ideas were influential. These propositions are concerned with the intellectual and contextual characteristics and structure of ideas. They differ conceptually from the neoliberal analysis of problem structure, which does not consider the degree to which certain ideas, because they can easily be communicated to the general public and to policy makers, lend themselves favorably to communicative action and therefore are taken up by policy entrepreneurs and used in mobilizing parties and stakeholders.

As was pointed out in chapter 1, the ideational approach to political analysis has recently become increasingly prominent in studies in comparative politics and international relations, and significant improvements have been made in understanding the role of ideas in policy development. But, despite these improvements, the independent impact of ideas is not always clearly documented in this literature. Rather than analyzing ideas in the abstract, it seems necessary that ideational studies better combine and link together ideas and actors. To further improve the ideational approach, it will be important to address more fundamental issues, particularly the following: What makes ideas significant? In what ways do ideas matter? And how are significant ideas transmitted or diffused?

To judge from this case, a combination of interest-based regime analysis and ideational analysis seems the most promising research strategy for future regime studies, at least for environmental policy and natural resources management studies. But it would be important to broaden the narrow focus on policy elites and government actors in regime studies. Regime scholars with a rationalistic bent mostly ignore mobilization of political and public support, justification of policy, and creation of political legitimacy in regime processes. Much research has focused on strategic action, instrumental bargaining, and rationalistic elements of elite behavior, and it would be useful to carefully examine the influence of widely accepted ideas and common interests.

It is misleading to consider narrow self-interest to be the only motivating force in international environmental politics. This study demonstrated the political weight of concerns for the public interest in policy initiation and regime development. Similarly, the importance of collective interests of states should be acknowledged. The interests of states are often at the same time converging and conflicting, and both integrative and distributive issues are at stake in regime processes. For instance, as this study illustrated, the transnational coalition of policy entrepreneurs instigated cooperation by influencing states' perception of the ocean dumping problem and their interests with respect to ocean dumping; however, it did not have a similar influence on states' perception of distributive issues existing outside this issue area that would be negotiated in the upcoming UNCLOS III. Perhaps surprisingly, nonenvironmental issues were among the most difficult ones when the regime was established. Neither did the transnational coalition shape or redefine how states perceived their basic economic interests.

Moreover, with respect to individual motivation, there is no reason to assume that politicians always are public spirited or that ENGOs do not try to achieve rewards and benefits for themselves, just as there is no reason to assume that politicians never are public spirited or that ENGOs are unwilling to receive private benefits. Regime analysis should reflect that conflict and cooperation, and private and public interest, are integrated, not essentially separable. The presence and significance of nonmaterial interests in regime processes should be acknowledged.

But is it possible to combine the prominent approaches to regime analysis in a fruitful way? It is encouraging that scholars recently have begun to examine this question.[11] In order to do so, however, closer attention should be paid to the scope of validity of power-based, interest-based, and knowledge-based regime theories.[12] Theorists mostly treat the three regime "schools" as if they are mutually exclusive, although in reality they deal with different aspects and different stages of regime establishment. To advance the present state of regime analysis, it will be necessary to identify with more precision the dependent variable, the path or paths to regime formation that theoretical models attempt to explain, and the aspects and stages with which they are primarily concerned. Until this is done, it will be rather futile to combine prominent regime theories.

Systematic attention to public ideas and public debate would add several dimensions to regime analysis. The ideational approach broadens the analysis to include the broader political arena, and state-society relationships are seen as important in understanding policy development. The intrinsic strength of this approach is that it connects three important factors in environmental cooperation: elites, policies (and their institutionalization), and the public realm. As repeatedly stressed, public opinion and support is considered of paramount importance in ideational analysis. Lack of public attention and support creates barriers and obstacles to attempts to build and change regimes. Education of the public is therefore often necessary before further policy initiatives can be undertaken. In the words of Majone (1989, p. 145), "major policy breakthroughs are possible only after public opinion has been conditioned to accept new ideas and new concepts of the public interest." It seems quite evident, for instance, that significant policy breakthroughs in the area of climate policy will not be possible until the American and the European general public is better informed about global warming and thus becomes more supportive of public policy targeted at this problem.[13]

Comparative studies of the influence of ideas on environmental and other policy areas would be valuable. What role do ideas play in environmental policy making in smaller states, developing countries, and regional groups, such as the European Union?[14] How are differences in domestic structure, political culture, and national regulatory style across countries shaping the ways in which ideas acquire influence over environmental policy making and, in turn, environmental regimes? Recent studies in the influence of ideas on macroeconomic policy, foreign policy, and human rights have concluded that different domestic political systems provide policy entrepreneurs, ideas, and norms with different opportunities and access points.[15] Such studies could provide a useful starting point when designing comparative studies on how ideas acquire influence over environmental policy making. There is also a need for developing and using testable theories of domestic politics in regime studies.[16]

The role of leadership in regime processes should also be examined more carefully. As noted, the case of the ocean dumping regime disproves Young's hypothesis that a regime arises only if two of the three forms of leadership suggested by him—structural, entrepreneurial, and intellectual—are provided. To develop hypotheses, it would be useful to compare the findings in this study with other cases where transnational coalitions of policy entrepreneurs control problem definition and agenda formation, mobilize stakeholder groups, craft solutions, and broker those solutions to powerful actors. In this way we would learn more about how various forms of leadership combine in regime processes. The role of ideational leadership, in particular, needs closer examination. It would also be important to identify with more precision under what conditions policy entrepreneurs are able to significantly influence international policy and interstate negotiations.[17] To what extent policy entrepreneurs tend to target some types of problems rather than others, and to what extent their strategies for influence vary according across problem types, should also be examined. This study suggests that policy entrepreneurs select problems to which it is possible to attract sufficient political and public attention, craft politically acceptable solutions, and broaden support. It would also be necessary to identify with more care the specific political resources controlled by individuals exerting influential leadership.

Chapter 9 suggested four primary roles for ENGOs in international environmental regimes. Comparative empirical research would be most useful to examine and further improve these hypotheses about ENGOs' sources of influence. Moreover, examining the role of ENGOs with respect to regimes would broaden the narrow focus on state and policy elites prevalent in mainstream literature on regimes.

Moreover, the 1993 global radwaste ban raises many broader questions: How autonomous are states when they formulate their environmental policy? How does the ability to accommodate organized environmental interests vary among states? To what extent can states pursue their interests in the face of political pressure, international public opinion, and a regime? These issues should be examined further. How epistemic communities might increase their influence on international policy development with respect to ENGOs should also be investigated. Similarly, it is relevant to examine how ENGOs use their technical competence in order to reach political objectives, and how they weigh their technical knowledge against their political beliefs. For instance, with respect to ocean dumping of waste, Greenpeace distrusted the scientific advice of marine scientists because the organization thought that it was in the scientists' professional interest to support ocean dumping (Stairs and Taylor 1992, pp. 122–127).

Finally, as discussed in detail, from the point of view of marine scientists and marine scientific expert groups, the global radwaste disposal ban is evidence of a *negative* regime development. Students of international politics acknowledge that regimes might have negative effects, but more attention should be paid to this important issue.[18] It is also relevant to ask whether ENGO influence should always be regarded as positive from an environmental protection and societal point of view.[19]

Looking Back and Looking Ahead

The global ocean dumping regime marks a significant step forward in nations' efforts to protect the oceans of the world. Thirty years ago some parts of the oceans were seriously damaged. It was increasingly realized that they could not absorb unlimited amounts of wastes safely. Competent national agencies, monitoring, keeping records, and licensing dumping permits, should together formulate policy for protection of the marine environment against dumping.

The radwaste disposal ban has been a stumbling block within the global ocean dumping regime. While some countries claimed victory, others reconsidered their membership. But it has also been a revelation. To all, this experience has illustrated that the world of science and technology is not one of safety, absolutes, and hard facts, but rather one of risks, probabilities, and uncertainty.[20] As chapter 8 showed, relying on science to provide conclusive and unambiguous scientific evidence of environmental damage has turned out to be much more difficult than the public and advocates of regulation imagined. Moreover, because radwaste disposal raised sensitive political and economical issues in many countries, decision making based solely on marine scientific knowledge was inadequate. It produced controversial, not consensual, policy.

The decision to globally ban radwaste disposal raises two significant issues in environmental management: the advantages of following a comprehensive approach and the inclusion of nonenvironmental concerns and issues in environmental decision making. A 1996 protocol to the London Convention shows that the members of the global ocean dumping regime, in addition to emphasizing precaution and prevention, indeed intend to pay more attention to these two issues in the future (LC 1997).[21]

Comprehensiveness was one of the key concepts of the strategy to protect the global environment laid out by the Stockholm secretariat in 1972. The secretariat was aware that decisions and institutions might be shifting problems into other sectors of the environment rather than coming to grips with them. As a global management arrangement for a single waste-management activity, the global ocean dumping regime could be shifting problems around rather than "solving" them. Wastes that are not disposed of at sea must ultimately be disposed of on land or in the air (by incineration).

In order to avoid transferring harm to other environmental sectors, the decision whether or not to dispose of at sea should, therefore, include a comparison with harm from using other disposal options.

The London Convention says, in its Annex 3, that the competent national agency should compare the risks from ocean disposal with the risks from land-based disposal methods before issuing a dumping permit. But such comparative risk assessments have not been carried out systematically. As was documented in chapters 8 and 9, the risks from land-based methods of disposal of low-level radioactive waste were not systematically taken into account when it was decided to ban radwaste disposal. The ban was largely based on the perceived risks to humans and the marine environment from ocean disposal.

Marine scientists agree that the ocean in principle has an assimilative capacity, and a consensus definition—"the amount of material that could be contained within a body of seawater without producing an unacceptable biological impact" (Stebbing 1992, p. 288)—was reached in 1979.[22] From this definition it follows that pollution is an unacceptable change to the environment but that change in itself does not constitute pollution. GESAMP and the Advisory Committee on Marine Pollution (ACMP) have endorsed the concept of assimilative capacity and thus distinguish between acceptable and unacceptable change to the marine environment.[23] In accordance with the Stockholm strategy, they stress that so-called holistic considerations should be made in all cases, radwaste disposal included (Bewers and Garrett 1987, p. 119). This, furthermore, would minimize the total harm inflicted on the environment.

As I discussed in chapter 9, GESAMP was greatly concerned over the ban on radwaste disposal. GESAMP viewed the ban as an expression of a lack of confidence in regulatory decision making concerned with issues characterized by scientific uncertain as well as a forerunner of the more recent trend within international environmental forums to adopt the precautionary principle. GESAMP also criticized the regulatory approach taken by most international marine pollution arrangements because "the occurrence or risk of pollution becomes the major criterion for regulatory action" (GESAMP 1991, p. 25). In its view, this is a conceptually flawed approach and it leads to haphazard regulation. GESAMP and ACMP instead stressed the need to distinguish between contamination and pollution and the importance of agreeing on a definition of acceptability with respect to environmental change.

The system of black and gray lists, essentially the global ocean dumping regime's approach to regulation, illustrates such shortcomings. It relies heavily on the categorization of harmful substances as either safe or unsafe. Marine scientists, however, do not categorize substances in this way, as they find that the "biological effects of toxic materials are a function of their concentration" (Stebbing 1992, p. 290). Put more crudely, it all depends on the dose. But the black and gray lists do not take into consideration the aggregate amount or concentration of wastes, the assimilative capacity of the receiving body of ocean water, various uses of the oceans (e.g. recreation, fishing, or exploitation of mineral resources), or dumping periods.[24] Furthermore, the convention, as one participant of early scientific working groups has noted, "give[s] no specific criteria for the inclusion of materials in [the black and gray lists] and . . . some of the substances were included on the basis of very little scientific evidence" (Norton 1981, p. 147). This regulatory approach of categorizing substances as either safe or unsafe for regulatory purposes must be changed if the global ocean dumping regime is to become more effective.

It now seems evident that much research is needed before marine scientists will know which contaminants can be assimilated, and in which quantities. Without comprehensive knowledge of the behavior of contaminants in the marine environment and their effects on humans, it will be impossible to make comparative risk assessments. Marine scientific research should consequently be supported. There will otherwise be no foundation for a more informed ocean dumping policy or for improvement of the black-and-gray-lists system. More knowledge will also have important implications for public legitimacy of policy. Since the debate on the concept of assimilative capacity presently is still evolving within the marine scientific community, the scientific basis of ocean protection policies is vulnerable to criticism.[25] A more permissive ocean dumping policy that lacked a firm scientific foundation would be difficult to legitimize.

The science-based principle originally adopted in the London Convention indirectly encourages manipulation of scientific and technical issues. It in addition drives economic interests underground. It will be important to create institutions for environmental protection that will take more than the scientific and technical aspects into account when making decisions with significant economic, employment, and social consequences for states. From this perspective, the 1985 resolution calling for studies of wider legal, social, economic, and political aspects, as well as scientific and technical issues, was a potentially useful step that might have paved the way for an improvement of the existing way of making decisions.

GESAMP and ACMP have stressed that overall regulatory priorities are in need of improvement and, on the administrative and regulatory level, that the aim is to spend the financial resources available for marine environmental protection cost-effectively. They have suggested the principle of justification and have emphasized that regulatory decisions concerning ocean dumping should maximize net benefits to society at large: "A justified practice will be one for which the combined benefits to the whole of society are considered to outweigh the combined deficits or detriment, environmental effects being only part of the latter." (Report of ICES Advisory Committee on Marine Pollution (Cooperative Research Report 167, ICES, Copenhagen, 1989), p. 124) In selecting the best option, therefore, many issues need to be addressed, including economic, social, political, and scientific considerations as well as the nature and extent of damage. Insufficient knowledge about various practices' potential for environmental damage and effects on humans, difficulties involved in estimating the societal and economic value of environmental quality, and other such issues could make use of this approach difficult. However, it addresses several of the shortcomings of the existing approach.

It will be a major challenge to develop a more comprehensive, holistic, and effective strategy for protection of the marine environment. A crucial point will be education of the public, politicians, and other decision makers about the health of the marine environment and risks to the marine environment and humans. The mass media should be used to educate the public and politicians when substances are not as hazardous as we previously thought. Though ecological disaster and threats to human health make good copy, new knowledge giving a more complete picture is being reported on in the mass media.

A major step in this direction may be taken by civil servants and government agencies educating the public, politicians, and the private sector about the choices that must be made between environmental protection and social and economic development, about the ultimate goal of minimizing the total harm inflicted on the environment, and about the need for research. It is also important to stimulate public deliberation about acceptability of risks and, more particularly, the costs and benefits of regulating minute risks.[26] ENGOs may also play a useful role in this regard, but sometimes it will be necessary that environmental groups and ultimately the "green public" reexamine their view of the balance between environmental protection and social and economic development. Nonetheless, as part of such a process, NGOs from the environmental community (which the public often trust more than they trust industry, government, and scientists) and from the private sector should be able to fully participate in the work of expert groups. Such groups might provide a context within which the participants can get acquainted with the environmental and health sciences' advances as well as with difficulties in detecting the impact of minute concentrations of substances and with the economic and social concerns that regulation must and should take into account. Also, in regard to radioactive matters, since the nuclear industry can hardly be said to have an impressive record of public trust, ENGOs can play an important role as nuclear watchdogs.[27] This is certainly in the interest of the public, and it might also be in the self-interest of the industry. Holistic alternatives to present regulation might also be developed more easily in a context that is less dominated by the traditions and protocol of international diplomacy. After several years' debate, it is encouraging that the 1991 consultative meeting unanimously agreed that ENGOs such as Greenpeace International could, in the future, participate in the work of the inter-governmental scientific working groups established under the global ocean dumping regime.

But it will be necessary to reduce the role of entrepreneurial politics if environmental policy goals and instruments are to be developed through public deliberation. To succeed, there would be a need for putting into place incentives or rewards—at least public and political recognition of the usefulness of deliberation—to those who stimulate deliberation. The substantive content of policy should figure prominently, and the symbolic aspects of policy should be de-emphasized. Although this alone might be difficult enough to achieve in the environmental field, special-interest groups should be expected to protect their interests and values vigorously. Moreover, sustained deliberation is unlikely to flourish in the international arena in situations such as that of radwaste disposal, which is characterized by an uneven distribution of benefits and costs across countries. In any case, to increase the effectiveness of environmental regimes, deliberation should certainly be stimulated whenever and wherever possible.

Widely accepted ideas and policy entrepreneurs can affect the deep normative structure of environmental regimes and environmental cooperation. Policy entrepreneurs who shape the principles, norms, and values that define particular issue areas have a significant influence on regimes. Their power stems from their ability to influence the way in which states and the international community conceive of environmental issues. Under certain intellectual and contextual conditions they influence environmental regimes by using ideas and metaphors strategically to focus public and political attention on issues, by broadening support behind their solutions, and by brokering among states and stakeholders. Transnational coalitions of policy entrepreneurs and ENGOs can perform crucial roles in such respects. They are potentially most influential when they are concerned with plus-sum issues. Unless they are able to manipulate state incentives, for instance by mobilizing domestic and international pressure through effective campaigning, they have little influence when issues are primarily distributive or redistributive.

It is a short step from interest-based analysis to prescriptive advice stressing the importance of designing negotiations and institutions that distribute the benefits and costs of cooperation evenly across states, make sure that the benefits of cooperation exceed costs, and that incentives to discourage free riding are provided. There is in a sense an equally short step to be taken by scholars of epistemic communities who claim that these communities help protecting the environment and increasing societal welfare more broadly.[28] But these theories might not be particularly helpful when developing policy prescriptions because, as I have demonstrated, neither the environmental and economic costs and benefits nor the scientific basis of popular policies necessarily attract as much attention as interest-based and knowledge-based regime analysis predict and hope.

However, ideas are no panacea. Just as improving scientific and technical knowledge does not necessarily increase cooperation, ideas do not necessarily facilitate regime formation or increase the effectiveness of environmental regimes. Ideas might be fundamentally wrong and lead to costly and unwise social decisions. As regime analysis is intended to be valuable to society and to policy practitioners and decision makers, it is necessary to examine the substantive contents of regime policies and show how ideas interact with interests in the rise and demise of regime policies. Building on such knowledge, it might be possible to correct socially harmful regime policies and create future environmental policies and regimes that increase aggregate welfare. Nonetheless, it seems quite likely that we in the future will witness other examples of international ideas and norms having significant effect on international policy, particularly on issues seen as threats to the environment and collective human welfare.

Key Events

1958 UN Conference on the Law of the Sea adopts resolution to control ocean dumping of radioactive waste.

1967 Nuclear Energy Agency of OECD begins oversight of ocean dumping of radioactive waste by European countries.

1969 UN surveys member states on desires for international regulation and reduction of ocean dumping.

1971 Intergovernmental Working Group on Marine Pollution meets for the first time.

1972 UN Conference on the Human Environment is held in Stockholm. London Convention is adopted.

1975 London Convention enters into force.

1979 Greenpeace begins campaign against ocean dumping of low-level radioactive waste in Atlantic.

1980 Japan announces intention to dump low-level waste in Pacific. United States considers resuming ocean dumping of low-level waste.

1983 London Convention adopts non-binding resolution calling for moratorium on ocean dumping of low-level radioactive waste pending completion of two-year expert review of technical and scientific aspects of such dumping.

1985 Results of the two-year scientific review are presented to London Convention contracting parties. Contracting parties adopt resolution calling for suspension of all ocean dumping of low-level radioactive wastes pending (1) a study of the political, legal, economic, and social aspects of dumping at sea, (2) an assessment of land-based options, and (3) an assessment of whether it can be proved that ocean dumping will not have negative impacts on human health and/or cause significant damage to marine environment.

1993 Results of study of political, legal, economic, and social aspects of ocean dumping and scientific and technical assessments of consequences of resuming such dumping are presented to the contracting parties' annual meeting. Contracting parties decide to prohibit ocean dumping of low-level radioactive waste.

Notes

Chapter 1

1. Reich 1990 is the seminal study of public ideas in the context of domestic public policy.

2. See also Risse-Kappen 1994, p. 209.

3. For social constructivism's notion of intersubjective beliefs and its rejection of the claim that all ideas are individual ideas or reducible to individual ideas, see Ruggie 1998a, pp. 16–22.

4. I use *transnational* to refer to regular interactions across national boundaries where at least one actor is a nonstate agent. For this definition, see Risse-Kappen 1995a, p. 3.

5. This number includes treaties that are "dead letters" and some that really are about issues primarily unconcerned with the environment.

6. For discussions of this and alternative definitions of international regimes, see Levy et al. 1995, pp. 270–274; Hasenclever et al. 1997, pp. 8–22. For a now-classic critique of this regime definition and of regime analysis more generally, see Strange 1983.

7. The question whether regimes "matter" was first raised in a prominent way by Krasner (1983, pp. 5–10).

8. To my knowledge, regime analysts have, apart from one brief mention, not paid attention to public ideas. For the exception, see Keohane 1997.

9. For a conceptual discussion of public value, as distinct from private value, see Moore 1995, pp. 27–56. For a comprehensive study of global public goods, see Kaul et al. 1999.

10. For this definition of regime change, see Krasner 1983, p. 4. Changes in rules and in decision-making procedures, but not in principles and norms, are changes *within* regimes.

11. I use the word 'good' to indicate that it is a normative question whether a policy is good or bad; the answer depends on the normative standard against which the policy is compared.

12. The Convention on the Prevention of Marine Pollution by Dumping of Wastes and Other Matter, signed in London on November 13, 1972, took effect on August 30, 1975 (*ILM* 11 (November 1972): 1291–1314). In 1992, governments decided to change the name by which this convention was generally known from "London Dumping Convention" to "London Convention," fearing that the convention otherwise could be understood as sanctioning ocean dumping.

13. "Nuclear Dumping Ban Voted," *Washington Post*, November 13, 1993. This statement reflects the view of the majority of governments.

14. For a discussion see Sabatier 1991, p. 149.

15. See, e.g., Hasenclever et al. 1997.

16. On reflectivism, see Keohane 1988. On cognitivists, see Haggard and Simmons 1987, pp. 509–513. For an influential contribution to constructivism, see Wendt 1992. Constructivism may be subdivided into individual schools of thought. For instance, Katzenstein et al. (1998) distinguish *conventional*, *critical*, and *postmodern* constructivism.

17. See Miles 1987, pp. 37–53; Spiller and Hayden 1988.

18. On the influence of ideas on international cooperation in public health, see Cooper 1989. On the influence of ideas on macro-economic policy, see Hall 1989a. On the influence of ideas on monetary policy, see Odell 1982. On the influence of ideas on trade policy, see Goldstein 1993. On the influence of ideas on development strategies, see Sikkink 1991. On the influence of ideas on human rights policies, see Sikkink 1993. On the influence of ideas on military strategy, see Mueller 1993. On the influence of ideas on foreign policy, see Risse-Kappen 1994. On the influence of ideas on international political change, see Checkel 1997. On the role of ideas in the EU integration process, see Risse-Kappen 1996.

19. The need to examine interactions among policy elites, ideas, and society more broadly is underlined in Jacobsen 1995. See also Checkel 1997, p. 131.

20. On ideational causation, see Ruggie 1998a, pp. 16–22.

21. For focusing events, see Kingdon 1984, pp. 99–105.

22. It is often necessary to discover possibilities for joint gains; they do not simply "exist." See, e.g., Sebenius 1992. See also Scharpf 1997, pp. 120–121.

23. For alternative definitions of public diplomacy, see Kremenyuk 1991, p. 24; Fortner 1994, pp. 34–35; Dobrynin 1995, pp. 532–535.

24. On process tracing and thick description, see King et al. 1994, pp. 36–41 and 226–228. On process tracing in studies of the policy effects of ideas, see Yee 1996, pp. 76–78.

25. My analysis draws heavily on Ringius 1997.

26. See also Tannenwald 1999.

27. For an analysis of the recent regime change in the Baltic Sea area, see Ringius 1996, pp. 23–38.

Chapter 2

1. Among non-ionizing types of radiation are microwave, ultraviolet, laser, and radio-frequency radiations.

2. For a discussion, see Templeton and Preston 1982, pp. 79–80.

3. Regarding the choice between disposal or storage of radwaste, Preston (1983, p. 108) writes: "Storage is expensive, and it may not in the long run offer any better chance than does prompt disposal in optimizing the choice between cost of protection from radiation and the value of the radiation detriment thus avoided. Storage is thus, except in a very few cases, no substitute for disposal, and disposal should be the preferred management option as soon as it offers a reasonable chance of optimizing with respect to the resulting radiation detriment." See also Rochlin 1979, pp. 95–96.

4. At least two American and one Soviet nuclear-powered submarines have been lost at sea, however. See National Advisory Committee on Oceans and Atmosphere 1984, p. 5.

5. The Treaty Banning Nuclear Weapon Tests in the Atmosphere, in Outer Space and Under Water of 1963, commonly known as the Limited Test Ban Treaty, has been signed by more than 100 states, including all the major powers except France and China. More than 2000 nuclear tests have been conducted around the world, 528 of them in the atmosphere. See Norris and Arkin 1998.

6. Bewers and Garrett (1987) add: "However, the mix of radioisotopes involved is different in each case and radioisotopes vary widely in the extent to which they can affect marine organisms and man, so that the total radioactivity is only a very rough guide to the risk. It must also be stated that the dumping cannot be considered safe just because the releases of radionuclides are small compared to the natural incidence of radionuclides in the environment."

7. One example is the *de minimis* risk approach, according to which a material may not be regarded as radioactive and therefore may be considered "below regulatory concern." For a discussion, see Whipple 1986, pp. 44–60.

8. Half-life: the time required for the decay of half of the radioactivity in a radioactive substance. Half-lives vary greatly; for example, iodine-131 has a half-life of 8 days, whereas plutonium-239 has a half-life of 24,000 years.

9. So far there are no permanent disposal facilities in the United States for high-level nuclear waste. In 1987 Congress directed the Department of Energy to determine whether Yucca Mountain in Nevada would be suitable as a permanent disposal facility for spent fuel and high-level radioactive waste.

10. On the early history of this disposal option, see Deese 1978.

11. See Carter 1987, pp. 11–14. According to Carter, containment of tailings need not create a major problem for a properly regulated industry. Others disagree; see, e.g., Goble 1983, pp. 170–171.

12. The Low-Level Radioactive Waste Policy Act—P.L. (96–573), December 22, 1980—is quoted from p. 85 of Parker 1988.

13. Information on the number of containers dumped and their radionuclide content is incomplete. A later evaluation of all available data for past U.S. dumping resulted in an increase in the estimate of the quantity of radioactivity dumped of about 25% from earlier estimates. See LDC 1985a, Annex 2, p. 12.

14. UN Convention on the High Seas, Geneva, Article 25, paragraph 1, quoted in "Note on International Conventions Relating to Radioactive Marine Pollution," *Nuclear Law Bulletin* 13 (April 1974), p. 41.

15. Ibid.

16. See also Finn 1983a, p. 71.

17. Harry Brynielsson chaired the panel. A not-for-publication version of the report is Radioactive Waste Disposal into the Sea—Report of the Ad Hoc Panel under the Chairmanship of Mr. H. Brynielsson (TO/HS/21, 6 April 1960).

18. See Finn 1983a, pp. 71–73.

19. European Nuclear Energy Agency, Radioactive Waste Disposal Operation Into the Atlantic-1967 (Paris: OECD, 1968), quoted on p. 12 of Dyer 1981.

20. See Preston 1983, p. 115.

21. See LDC 1985a, Annex 2, p. 73.

22. Hagen (1983, p. 49) estimates the radioactivity to be 4.3×1015 Bq. The information on the number of containers dumped and their radionuclide content is incomplete.

23. Hagen (1983, p. 51), who also assesses the number of containers dumped at about 350, estimates the activity to be 8.5×1012 Bq. This is practically identical to the activity reported by Holcomb.

24. Eighty countries participated: Afghanistan, Argentina, Australia, Austria, Bahrain, Bangladesh, Barbados, Belgium, Bolivia, Brazil, Britain, Byelorussian SSR, Cameroon, Canada, Chile, Denmark, Dominican Republic, Egypt, El Salvador, Ethiopia, Federal Republic of Germany, Fiji, Finland, France, Gambia, Ghana, Greece, Guatemala, Haiti, Honduras, Iceland, India, Indonesia, Iran, Ireland, Italy, Ivory Coast, Jamaica, Japan, Jordan, Kenya, Korea, Kuwait, Liberia, Malaysia, Mexico, Monaco, Morocco, Nepal, Netherlands, New Zealand, Nicaragua, Nigeria, Norway, Pakistan, Panama, Paraguay, Philippines, Portugal, San Marino, Saudi Arabia, Senegal, Somali, South Africa, Spain, Sri Lanka, Sweden, Switzerland, Thailand, Tonga, Trinidad and Tobago, Tunisia, Uganda, Ukrainian SSR, United States, Uruguay, Soviet Union, Venezuela, Yemen, Zambia. Twelve governments sent observers.

25. However, Switzerland did not participate in 1973, and Belgium did not participate in 1974 or in 1977.

26. See Van Dyke 1988, p. 86. According to another source, up to a million drums, with an annual radioactivity of approximately 10^5 Ci during the operational phase, would be disposed of over the decade. See Finn 1983b, p. 215. Japan was of the opinion that the global dumping regime justified its planned low-level dumping. See Van Dyke et al. 1984, p. 743; Branch 1984, p. 327.

27. Protest was rising against this dumping.

28. Principally Greenpeace,

29. South Pacific island nations took the lead.

30. "United Kingdom: Ocean Disposal Operations to Continue," *Nuclear News* (July 1983), p. 50.

31. "Ocean Disposal: Japan Calls a Halt," *Nuclear News* (February 1985), p. 118.

32. LDC 1984, p. 31.

33. LDC 1985a, Annex 2, p. 71.

34. In October 1987, for example, the Japanese delegation to that year's meeting of the global ocean dumping regime stated that "although [Japan] is not presently dumping radioactive wastes at sea, it regards sea dumping as an important option for the future" (Van Dyke 1988, p. 82).

35. See, e.g., Deere-Jones 1991, pp. 18–23.

36. Personal communication, National Agency of Environmental Protection, Denmark, October 1988. Nordic environment ministers have recently appealed to British authorities to halt radioactive discharges from Sellafield. See "Stop Effluent from Sellafield," *Norden This Week*, April 1998.

37. See Sielen 1988.

38. See Fairhall 1989; Pienaar 1989; Jones 1989, p. 251. On the recent controversy in Britain over the storage of decommissioned nuclear submarines, see Gray 1996.

39. The ban does not apply to waste containing de minimis levels of radioactivity. The members will conduct scientific reviews of the ban 25 years after it goes into force and at 25-year intervals thereafter.

40. E. Haas (1980, p. 386) defines regime stability in an almost identical way.

41. See, e.g., Sand 1992.

42. See LDC 1985b, p. 10.

43. See IMO 2000.

44. See Timagenis 1980, pp. 4–9.

Chapter 3

1. Despite many insights and stimulating propositions regarding norms, law, order, states, and international organizations, leading ideational studies generally are more concerned with structure than with transnational and nonstate actors and agency more specifically. See Kratochwil 1989; Franck 1990; Onuf 1989. The literature concerned with ideas-based policy change and constructivist literature on international norms has rightly been criticized for neglecting agency. See, e.g., Checkel 1999. The current debate about how to conceptualize the interrelationships between states and the international system, as well as the consequences for our understanding of international relations that follow from adopting various conceptualizations, was introduced in Wendt 1987.

2. On policy entrepreneurs in American politics, see Kingdon 1984 and Polsby 1984. On policy entrepreneurship by EU officials, see Majone 1996a, especially chapter 4.

3. Middle-range theories are often used in environmental studies, ideational analyses, and regime analyses. See, e.g., Liefferink 1996, pp. 40–57; Checkel 1997, pp. 3–12; Stokke 1997, pp. 27–63.

4. Major realist works include Carr 1939, Morgenthau 1948, Aron 1966, Waltz 1954, Waltz 1979, and Gilpin 1981. For the recent debate, see Kegley 1995.

5. On interest-based theory and reflective theory, see Keohane 1988. On neorealism and neoliberalism, see Nye 1988. For the recent debate, see Baldwin 1993. For the most widely quoted work on collective action problems, see Olson 1965. To analyze international cooperation, especially how and when public goods might be provided, interest-based analysis often draws on economists' studies. For a good general discussion, see Miller 1997.

6. Modern reflective or liberal theory, like its predecessors "functionalism" and "neofunctionalism," stresses domestic society's impact on international society, interdependence, and international institutions. Major liberal works include Deutsch et al. 1957 and E. Haas 1958. For the classic outline of functionalistic international theory, see Mitrany 1966. E. Haas (1980, 1983, 1990) has extensively examined the cognitive and philosophical foundation of international cooperation. For empirical work in this tradition, see Haas et al. 1977.

7. See P. Haas 1990b, p. 348.

8. For comparable cases, see George and McKeown 1985, pp. 21–58.

9. See, e.g., Orren 1990, pp. 22–24.

10. Inattention to the views and influence of mass publics is quite widespread among international politics scholars, including a number of recent ideational studies in international politics. For a discussion, see Jacobsen 1995.

11. This definition comes close to how Kingdon (1984, pp. 129–130, 188–193, 214–215) defines policy entrepreneurs. For a promising systematic approach drawing on both political and economic theory, see Schneider and Teske 1992.

12. On policy windows, see Kingdon 1984, pp. 173–204.

13. On de facto natural coalitions, see Sebenius 1992, pp. 352–353.

14. The seminal article on this metaphor for domestic-international interaction is Putnam 1988.

15. Furthermore, Kelman (1990, p. 40) notes parenthetically, "the organization of environmentalists into interest groups generally *followed* environmental legislation, rather than preceded it."

16. See Majone 1989, pp. 35–36.

17. For examples, see Kingdon 1984, p. 189.

18. The terms *issue network*, *policy community*, and *policy subsystems* have been used to describe informal networks of policy specialists active in setting agendas and formulating policies. On issue networks, see Heclo 1978, pp. 87–124. On policy communities, see Kingdon 1984, pp. 122–128. On policy subsystems, see Sabatier 1988, p. 131.

19. On political feasibility, see, e.g., Majone 1975 and Underdal 1991.

20. Litfin (1994, p. 30) similarly posits that the influence of a trans-scientific discourse, in addition to the authority of its agents, depends upon the "contextual and substantive constitution of discourse."

21. As an example of cognitive studies underscoring the need for simplicity, see Odell 1982, p. 68.

22. For similar propositions regarding policy makers selecting among economic ideas, see Woods 1995, pp. 172–173.

23. On "the ozone hole," see Litfin 1994. On the slogan "Save the whales," see Andresen 1989. Similar to his fellow biologists, the internationally renowned scientist Edward O. Wilson used the term *biological diversity* before the issue became a prominent international issue in the late 1980s. But a few staff members of the U.S. National Academy of Science persuaded him to use the term *biodiversity* in the campaign for preservation of biological diversity because the term is "simpler and more distinctive . . . so the public will remember it more easily." The issue soon attracted increasing international attention, and a treaty, the Convention on Biological Diversity, was signed by more than 150 governments in Rio de Janeiro in June of 1992. "The subject," wrote Wilson (1994, pp. 359–360), "surely needs all the attention we can attract to it, and as quickly as possible." The metaphor of the Earth as a spaceship on which humanity travels, dependent upon its vulnerable supplies of air and soil, first appeared in a speech given by the U.S. ambassador to the United Nations before the UN Economic and Social Council in 1965. The prominent economist and ecologist Barbara Ward had drafted this speech, and she used *Spaceship Earth* as the title of a book published in 1966.

24. See, e.g., Conlan et al. 1995, p. 135.

25. Jönsson (1993, pp. 209–213) discusses how metaphors might influence the perception, choices, and strategies of decision makers involved in regime creation, but not their effects on the public.

26. Goldstein (1993, pp. 12, 255–256) underlines that ideas need to fit with underlying ideas and social values in order to gain support among political entrepreneurs and the attentive public. On the notion of the fit of ideas, see Yee 1996, pp. 90–92. In a similar way, the norms-oriented literature stresses adjacency, precedent, and fit between new and old norms. See, e.g., Finnemore and Sikkink 1998, p. 908. Jacobsen (1995, p. 304) stresses the importance of ideas' satisfying minimum societal standards of distributive fairness.

27. For a similar conclusion in the context of intergovernmental negotiations, see Garrett and Weingast 1993, p. 186.

28. Scharpf 1997 (pp. 130–135) also emphasizes the role of "arguing" in increasing aggregate welfare through joint problem solving. For a recent analysis of the role of argument and persuasion in international relations, see Risse 2000.

29. Waltz concludes, in a note, that "states [in such situations] face a 'prisoners' dilemma,'" and that "if each of two parties follows his own interest, both end up worse off than if each acted to achieve joint interests" (1986, p. 129).

30. On the tragedy of the commons, see Ostrom 1990, pp. 2–3.

31. For example, an official from the Jamaican Ministry of Foreign Affairs has noted that "in developing countries, the conflicts ["between any obligation to prevent pollution of the marine environment and the effects of implementing that obligation on economic development"] are far more intense because the options open for economic development are limited" (Kirton 1977, p. 280).

32. Grieco (1990, p. 10) has emphasized that states are primarily concerned about their physical survival and their political independence. Snidal (1991) has challenged the realist claim that states worry about how well they fare compared to other states (relative gains), and not simply how well they fare themselves (absolute gains).

33. For example, in April of 1977 the Carter administration decided to defer commercial reprocessing of nuclear waste indefinitely because of the feared risk of proliferation of nuclear weapons. See Power 1979.

34. See Keohane 1980, pp. 131–162.

35. On benevolent and coercive hegemonic leaders, see Hasenclever et al. 1997, pp. 90–95.

36. Grieco (1990, p. 10) has emphasized the Waltzian view on cooperation claiming that such bargains would not take place: ". . . a relative-gains problem for cooperation [exists]: a state will decline to join, will leave, or will sharply limit its commitment to a cooperative arrangement if it believes that gaps in otherwise mutually positive gains favor partners." For a recent discussion, see Snidal 1991 and Powell 1991. For a critical discussion of Grieco's view, see Milner 1992.

37. For the definition of mixed-motive games, see Schelling 1960, p. 89.

38. See, e.g., Keohane's discussion of bounded rationality and incomplete and partially available information (1984, pp. 110–116).

39. On the games Prisoners' Dilemma, Stag Hunt, and Chicken, see Oye 1986, pp. 1–24. See also Axelrod and Keohane 1986, pp. 228–232.

40. Realists, however, doubt that the "shadow of the future" necessarily makes cooperation more likely. See Grieco 1990, p. 227.

41. On N-person games, see Snidal 1986, pp. 52–55. On the significance of the number of actors, see Snidal 1995.

42. On the benevolent effects of institutions' transforming N-person games into collections of two-person games, see Axelrod and Keohane 1986, pp. 237–239.

43. On issue density, see Keohane 1984, pp. 79–80.

44. See, e.g., Keohane 1984, pp. 78–80. Those who stress that hegemons act on selfish reasons when they provide public goods build upon Olson's (1965, pp. 33–34) argument about "privileged groups."

45. Keohane's (1984) functional theory of international regimes explains regimes as means at the disposal of states wanting to cooperate. On functional explanations, see Elster 1983, pp. 49–68.

46. On focal points, see Schelling 1960, p. 70.

47. See Young 1991.

48. See Young 1989, pp. 373–374; Young 1991, pp. 302–305.

49. See Russett and Sullivan 1971, pp. 853–854.

50. The literature on how perception and misperception at the level of individuals and groups influence foreign policy making is not included in the reflective group. See, e.g., Jervis 1976; Steinbruner 1974.

51. More recently, E. Haas (1990, p. 12) has concluded that "when knowledge becomes consensual, we ought to expect politicians to use it in helping them define their interests."

52. P. Haas (1990b, p. 349) claims that environmental regime formation is not dependent upon states, even prominent ones, providing leadership.

53. For an example of how ideological differences among decision makers matter, see Odell 1982, p. 358.

54. See, e.g., Ruggie 1983, pp. 195–231.

55. See also Haggard and Simmons 1987, pp. 509–513.

56. E. Haas (1990, p. 42) concludes that "the success of an epistemic community thus depends on two features: (1) the claim to truth being advanced must be more persuasive to the dominant political decision makers than some other claim, and (2) a successful alliance must be made with the dominant political coalition.... Epistemic communities seek to monopolize access to strategic decision-making positions."

57. It seems beyond doubt that Cousteau framed this regional environmental problem and attracted attention to it. In the words of the first editor of the scientific journal *Marine Pollution Bulletin*: "Twenty years ago, the Mediterranean had a thoroughly bad reputation for its polluted waters. Jacques Cousteau could say that the Mediterranean was 90% dead and, whatever that might have meant, it highlighted an undoubtedly serious state of affairs.... In these circumstances, the Mediterranean was an obvious target for attention by the United Nations Environmental Programme under its Regional Seas Programme." (Clark 1989b, pp. 369–370)

58. On the international political impact of the 1967 *Torrey Canyon* oil spill between France and Britain, see M'Gonigle and Zacher 1979, pp. 143–147. On the impact of the discovery of a seasonally depleted ozone layer over Antarctica, see Litfin 1994. On the impact of Chernobyl on regulatory approaches, see Liberatore 1999.

Chapter 4

1. For a brief account of the asbestos case, see Sapolsky 1990, pp. 83–96. For a critical account of government regulation of the insecticide DDT, see Wildavsky 1997, pp. 55–80.

2. For a study of nuclear fears over time, see Weart 1988.

3. For an early article on the pollution of Lake Erie, see Powers and Robertson 1966. According to Hasler and Ingersoll (1968, p. 153), "it took the visual (and olfactory) impact of a huge body of water, Lake Erie, suffocating as a sump for industrial waste, sewage, and urban and rural runoff to bring the problem of water pollution dramatically to the public eye," and "some now pronounce Lake Erie 'dead.'"

4. Barry Commoner devoted a chapter of *The Closing Circle* (1971) to the subject. For a rejection of Commoner's view, see Maddox 1972. For denials of the poor condition of Lake Erie, see Grayson and Shepard 1973. Since 1972, according to mass media reports, the state of the Great Lakes seemed to be improving. See Goth 1974 and Tunley 1974.

5. For Nixon's State of the Union message, see *Congressional Quarterly: Weekly Report*, January 23, 1970, pp. 245–248. See also "Nixon's Spirit of '76," *Newsweek*, February 2, 1970; "Pollution: The Battle Plan," *Newsweek*, February 23, 1970.

6. Senator Henry Jackson had proposed a permanent three-member White House Council on Environmental Quality. It was established by a law which originated in legislation introduced by Representative John Dingell, chairman of the House Subcommittee on Fisheries and Wildlife Conservation. For a brief discussion of environmentalism among American presidents, see Nobile and Deedy 1972, pp. xiii–xvii.

7. According to Moore (1990, p. 67) and Marcus (1980, p. 287), Nixon saw the environment issue as "phony" and "faddish."

8. For Nixon's message to Congress on pollution of the Great Lakes, see "Great Lakes Pollution," *Congressional Quarterly: Weekly Report*, April 17, 1970, pp. 1050–1051.

9. "Don't Go Near the Water," *Newsweek*, October 19, 1970. A few months later, *Newsweek* listed as America's ten "filthiest" rivers the Ohio, the Houston Ship Channel, the Cuyahoga, the Rouge, the Buffalo, the Passaic, the Arthur Kill, the Merrimack, the Androscoggin, and the Escambia ("From Sea to Shining Sea," January 26, 1970).

10. Gordon J. F. MacDonald, "Statement," in U.S. Senate, Ocean Waste Disposal: March 2, 3; April 15, 21, 22, and 28, 1971 (92nd Congress, 1st Session), p. 153

11. See, e.g., John D. Parkhurst, "Statement," in U.S. Senate, Ocean Waste Disposal: March 2, 3; April 15, 21, 22, and 28, 1971 (92nd Congress, 1st Session), 1971, p. 124.

12. For the brief mention in the House, see *Congressional Record: House*, September 8, 1971, p. 30863.

13. "International Aspects of the 1971 Environmental Program: Message from President Nixon to the Congress," *Department of State Bulletin*, March 1, 1971, p. 254.

14. Ibid., p. 255. The CEQ's recommendation with respect to international cooperation (CEQ 1970, p. v) said: "Finally, this report recognizes the international character of ocean dumping. Unilateral action by the United States can deal with only a part-although an important part-of the problem. Effective international action will be necessary if damage to the marine environment from ocean dumping is to be averted."

15. "Marine Protection, Research, and Sanctuaries Act of 1972," section 109; quoted from Senate Report no. 451, Marine Protection, Research, and Sanctuaries Act of 1972 (92nd Congress, 1st session) (1971), p. 4272.

16. The terms *assimilative capacity*, *environmental capacity*, *accommodative capacity*, and *absorptive capacity* are used to identify the capacity of the oceans to receive wastes safely.

17. On the influence of the assimilative capacity concept, and for doubts expressed about the concept, see Spiller and Rieser 1986.

18. For a balanced assessment of known and unknowns about marine pollution, see Knauss 1973, pp. 313–328.

19. In his "Message on the Environment" (*Congressional Quarterly: Weekly Report*, February 13, 1970, p. 436), Nixon said: "I propose that State-Federal water quality standards be amended to impose precise effluent requirements on all industrial and municipal sources. These should be imposed on an expeditious timetable, with the limit for each based on a fair allocation of the total capacity of the waterway to absorb the user's particular kind of waste without becoming polluted."

20. U.S. House of Representatives, Subcommittee on Fisheries and Wildlife Conservation and the Subcommittee on Oceanography of the Committee on Merchant Marine and Fisheries, Ocean Dumping of Waste Materials: April 5, 6, 7, 1971 (92nd Congress, 1st session), p. 454.

21. For the EPA's view of the assimilative capacity concept, see William Ruckelshaus, "Statement," U.S. House of Representatives, Subcommittee on Fisheries and Wildlife Conservation and the Subcommittee on Oceanography of the Committee on Merchant Marine and Fisheries, Ocean Dumping of Waste Materials: April 5, 6, 7, 1971 (92nd Congress, 1st session), p. 454. Ruckelshaus criticized the "false premise" that the oceans have an unlimited assimilative capacity but seemed positive toward ocean disposal if there was a clear "benefit" to the ocean. Compare *Congressional Record: Senate*, March 16, 1971, p. 6572, and "Statement," in U.S. Senate, Ocean Waste Disposal: March 2, 3; April 15, 21, 22, and 28, 1971 (92nd Congress, 1st Session), p. 280. Neither did the CEQ favor a ban on all dumping; see MacDonald, "Statement," in U.S. House of Representatives, Ocean Waste Disposal, pp. 146–148.

22. Ruckelshaus concurred with this view. See Ocean Dumping of Waste Materials, p. 451.

23. For a good discussion of these scientists' values (which, however, overlooks that politicians' values influenced whether the scientific and expert advice given was transformed into public policy, and which also ignores the public view of the acceptability and desirability of ocean dumping), see Spiller and Rieser 1986.

24. David D. Smith, "Statement," U.S. Senate, Ocean Waste Disposal: March 2, 3; April 15, 21, 22, and 28, 1971 (92nd Congress, 1st Session), p. 206.

25. On the lack of knowledge of effects of wastes on the ocean environment, and the possible benefits of waste disposal, see, respectively, Parkhurst, "Statement," in U.S. Senate, Ocean Waste Disposal: March 2, 3; April 15, 21, 22, and 28, 1971 (92nd Congress, 1st Session), 1971; Ralph Porges, "Statement," in U.S. Senate, *Ocean Waste Disposal: March 2, 3; April 15, 21, 22, and 28, 1971* (92nd Congress, 1st Session), pp. 122–142.

26. Senate Report no. 451, Marine Protection, Research, and Sanctuaries Act of 1972 (92nd Congress, 1st session), 4239–4240.

27. See P. Haas 1990a, p. 74. For several examples of rivalry and competition among international organizations involved in issues of marine pollution, see Boxer 1982. On how the IMCO (then the IMO) made use of the *Torrey Canyon* oil spill to invigorate itself, see M'Gonigle and Zacher 1979, p. 42.

28. See also other contributions in this special issue of *International Organization* on the UN and the environment (volume 26, spring 1972).

29. Christian A. Herter Jr., "Statement," in U.S. Senate, Committee on Commerce, and House of Representatives, Committee on Science and Astronautics, International Environmental Science: May 25 and 26, 1971 (92nd Congress, 1st session), p. 27.

30. For example, in a 1974 internal memo (quoted in Haas 1990a, p. 256), a high-level UNEP official wrote: "I continue to support FAO's tactic of stressing the effect on fisheries as a way to stimulate action on pollution in the Mediterranean, but suspect that the tourist angle is also one which should be played, particularly in relation to oil on beaches, but also in relation to sewerage."

31. On the preparations for the conference, see Farvar and Milton 1972, pp. 974–975.

32. See Caldwell 1970, pp. 198–204; Caldwell 1984, pp. 39–40.

33. Pryor was quoting from what he called Paul Ehrlich's "persuasive fable on the future of the ocean titled *Eco-Catastrophe*."

34. PCBs are chlorinated organic substances that are very toxic and do not degrade easily. They accumulate in organisms and move up the food chain. Among their effects are reduced reproductive ability and reduced resistance to infection.

35. "Interdependency Resolution: A Committee Resolution Prepared by James Pepper and Cynthia Wayburn," in Johnson 1970, p. 337. See also Flateboe 1970, pp. 179–182. Barry Commoner also recommended that action should be taken through the United Nations in his background paper "The Ecological Facts of Life," which was reprinted in an anthology of papers presented at the conference. (For an earlier version of the latter paper, see Commoner 1970.)

36. In 1966 the Subcommittee on Marine Science and Its Application (formed by the Administrative Committee on Co-ordination of the United Nations) and IMCO surveyed all UN member countries on ocean pollution. Out of the 66 replies, 19 governments declared that pollution of the seas was no problem; of the 47 stating positively that pollution was a problem, 36 indicated that harm to living resources was a major concern, 20 identified threats to human health as a concern, 12 referred to hindrance of marine activities, and 19 to reduction of amenities. See Pravdic 1981, pp. 3–4.

Chapter 5

1. Committee on International Environmental Affairs, Task Force III, US Priority Interests in the Environmental Activities of International Organizations (1970), p. 23.

2. See note 8 below.

3. Some have claimed that the global ocean dumping regime was created in direct response to these dumpings (Barston and Birnie 1980, p. 113). For international reactions, see Joesten 1969, p. 152; Böhme 1972, p. 98.

4. One congressman said during the House debate on the ocean dumping bill: "That emergency situation demonstrated that we had virtually no national policy or means of control for ocean dumping, and we had to stand by and watch the Army dump nerve gas into the Atlantic Ocean off the coast of Florida." (*Congressional Record: House*, September 9, 1971, p. 31154) See also *Congressional Record: House*, October 16, 1973, p. 34298.

5. For international responses to the Army dumpings, see Friedheim 1975, pp. 173–174. This incident demonstrated to lawyers the inadequacy of international law in this area; see Schachter and Serwer 1971, pp. 107–108.

6. See Smith 1970.

7. Rep. Alton Lennon, Rep. John Dingell, and Sen. Ernest Hollings (chairman of the Senate Subcommittee on oceans and atmosphere) were the driving forces behind U.S. ocean dumping legislation. On Hollings's strong interest in marine science and technology, see Gillette 1972, pp. 729–730. Lennon and Dingell had for several years prior to 1971 been concerned with the degradation of the marine environment and were largely responsible for the development of the National Environmental Policy Act and the establishment of the CEQ; see *Congressional Record: House*, September 8, 1971, p. 30854.

8. During the Senate debate on the ocean dumping bill, Hollings said: ". . . the actual goal of trying within 5 years to set a policy against dumping the committee will welcome and gladly go along with" (*Congressional Record: Senate*, November 24, 1971, p. 43068). See also Lumsdaine 1976, pp. 771–772; Bakalian 1984, p. 213. This group's continued interest in minimizing all ocean dumping is evident from the following exchange between Hollings and Mr. Rhett, a spokesman for the EPA, at the 1975 Oversight Hearing for the dumping act. Hollings: "If you had to make the choice from your vantage point, would you ever choose the ocean over a land site? I mean, I have been listening to this testimony about all the progress in phasing dumping out, and we're starting here and there, and now you act like you're going to start up something that never was." Rhett: "Let's say you have no heavy metal contaminants or anything of this nature and no land available for disposal. I'm not sure. Maybe it is better to burn it and pollute the air, but I think that we should evaluate all methods. I am not saying that it should be in the ocean, but, I am saying that I think all methods of disposal should be considered." Hollings: "We looked at all methods of disposal and we looked at oceans. . . . We made that determination. We are not looking around to find places to dump. . . . You guys had better stay in that one direction because we'll amend the law to make sure you do." Rhett: "I think we are, but I do not believe that the act, as such, precludes ocean dumping. It says 'regulate'. Hollings: "Well, we'll look at that and make sure because we want to go in one direction on this one. We're trying to clean up the oceans. Go right ahead." (quoted in Lumsdaine 1976, p. 772)

9. See speech by Robert M. White, Administrator-Designate of NOAA, to American Oceanic Organization, February 4, 1971 (*Congressional Record: Senate*, February 4, 1971, pp. 1670–1672).

10. See also Sen. William Roth Jr. in *Congressional Record: Senate* (April 1, 1971), p. 9209; Rep. Dante Fascell, "Statement," in U.S. House of Representatives, Ocean Dumping of Waste Materials, pp. 137–141; Sen. Ernest Hollings, *Congressional Record: Senate*, May 12, 1971, p. 14667.

11. Sen. Ernest Hollings, in U.S. Senate, Subcommittee on Oceans and Atmosphere of the Committee on Commerce, International Conference on Ocean Pollution: October 18 and November 8, 1971 (92nd Congress, 2nd session), p. 40.

12. Sen. William Roth Jr., letter to President Nixon, October 13, 1970, reprinted in *Congressional Record: Senate*, April 1, 1971, p. 9209.

13. In addition to a number of unidentified representatives of foreign governments, the ambassadors of Spain, South Africa, Honduras, Barbados, and Portugal were present.

14. Scott Carpenter, Jacques Cousteau, Christian Herter, Mark Morton, Barry Commoner, Hugh Downs, and Thor Heyerdahl participated in this conference.

15. U.S. House of Representatives, Subcommittee on Fisheries and Wildlife Conservation and the Subcommittee on Oceanography of the Committee on Merchant Marine and Fisheries, Ocean Dumping of Waste Materials: April 5, 6, 7, 1971 (92nd Congress, 1st session), p. 163. On the planned French dumping in the Mediterranean, see "Atomic Disposal Alarms Riviera," *New York Times*, October 11, 1960; "Riviera Resorts Threaten Strike," *New York Times*, October 12, 1960; and "France to Delay Atomic Disposal," *New York Times*, October 13, 1960. On Cousteau's involvement in the French anti-nuclear movement, see Nelkin and Pollak 1981, p. 91.

16. See Schachter and Serwer 1971, p. 90. The Joint Group of Experts on the Scientific Aspects of Marine Pollution (GESAMP) was established by several specialized agencies of the United Nations in 1969.

17. Sen. Ernest Hollings, in U.S. Senate, Subcommittee on Oceans and Atmosphere of the Committee on Commerce, International Conference on Ocean Pollution: October 18 and November 8, 1971 (92nd Congress, 2nd session), p. 40.

18. On Barry Commoner, see Goodell 1975, pp. 60–69.

19. Hollings, International Conference on Ocean Pollution, p. 71.

20. International Conference on Ocean Pollution, p. 126.

21. See *Congressional Record: House*, November 9, 1971, pp. 40233–40235. Many statements given by Heyerdahl at U.S. hearings and the Stockholm conference were reprinted in the *Congressional Record*.

22. The *New York Times* had an apocalyptic editorial line. A few weeks earlier (October 24), its Sunday magazine had published an article by Michael Harwood titled "We Are Killing the Sea Around Us." The latter article's daunting conclusions were characteristic of the time: "'Lake Erie may or may not be restored within 50 years,' Dr. Max Blumer of Woods Hole wrote last December, 'but a polluted ocean will remain irreversibly damaged for many generations.' And who would care to argue that a dead ocean would not mean a dead planet?"

23. See, e.g., Christenson 1976, p. 2.

24. See also McManus 1982, p. 90; Clark 1989b, pp. 369–372. In a 1978 fund-raising letter, in which no numbers were given, Cousteau repeated the frightening possibility of the death of the oceans: "I beg you not to dismiss this possibility [of death of the oceans] as science fiction. The ocean can die, these horrors could happen. And there would be no place to hide!" (quoted in Douglas and Wildavsky 1982, p. 170)

25. House Report no. 361, Marine Protection, Research, and Sanctuaries Act of 1971 (92nd Congress, 1st session) (1971), p. 14.

26. Ibid., p. 12.

27. On the Ehrlichs, see Goodell 1975, pp. 11–18. For a critical view of Paul Ehrlich, see Efron 1984, pp. 33–35.

28. House Report no. 361, pp. 11–12. (This quotation was taken from Paul R. Ehrlich and Anne H. Ehrlich, "The Food-from-the-Sea Myth," *Saturday Review*, April 4, 1970.) For the House debate on the committee report, see *Congressional Record: House*, September 8, 1971, pp. 30850–30866.

29. Lawrence Coughlin, "Statement," in U.S. House of Representatives, Ocean Dumping of Waste Materials, p. 161.

30. House Report no. 361, p. 12. For the full letter, see U.S. House of Representatives, *Ocean Dumping of Waste Materials,* pp. 161–162.

31. Ibid., p. 12.

32. James T. Ramey, "Statement," in U.S. House of Representatives, Ocean Dumping of Waste Materials, p. 264. See also CEQ chairman Russell Train, "Statement," in Ocean Dumping of Waste Materials, pp. 205–207.

33. U.S. House of Representatives, Ocean Dumping of Waste Materials, p. 375.

34. See *Congressional Record: Senate*, September 21, 1972, p. 31700; October 3, 1972, p. 33307; October 13, 1972, p. 35842.

35. See Pearson 1975a, pp. 207–219.

36. See, e.g., Rep. Lawrence Coughlin, "Statement," in U.S. House of Representatives, Ocean Dumping of Waste Materials, p. 161.

37. See *Congressional Record: House*, September 9, 1971, pp. 31151 and 31354. This was also repeated during the hearings on ocean dumping; see Rep. Dante B. Fascell, "Statement," in U.S. House of Representatives, Ocean Dumping of Waste Materials, p. 139. According to a three-paragraph article printed in the *Washington Post* of September 17, 1970, and reprinted in *Congressional Record: Senate*, October 6, 1970, p. 35134: "'The oceans are dying,' Cousteau says. 'The pollution is general.' That's the appraisal of Jacques Yves Cousteau, back from 3½ years' exploration and moviemaking around the world. 'People don't realize that all pollution goes to the seas. The earth is less polluted. It is washed by the rain which carries everything into the oceans where life has diminished by 40 per cent in 20 years,' the underwater explorer said."

38. See *Congressional Record: House*, September 9, 1971, pp. 31129–31160.

39. One congressman said: "The general public feeling is that government is unresponsive and that the individual is powerless to affect his environment. That is not the case when legislation such as this is enacted." (*Congressional Record: House*,

September 9, 1971, p. 31155) Another congressman said during the House debate on the dumping bill: "We can afford to wait no longer. We must pass this bill. We must demonstrate to the American people that Congress is ready, willing-and, yes, able-to act in this area of critical need. Let us not delay." (*Congressional Record: House*, September 8, 1971, pp. 30859–30860)

40. The House passed the ocean dumping bill by a vote of 305 to 3. The Senate passed the ocean dumping bill by a vote of 73 to 0.

41. On the debate in the House, see letter from the Sierra Club, Friends of the Earth, and other ENGOs, *Congressional Record: House*, September 8, 1971, p. 30853. See also *Congressional Record: House*, September 9, 1971, pp. 31151–31152. On the debate in the Senate, see *Congressional Record: Senate*, November 24, 1971, p. 43060.

42. See also Newman 1973. The newspapers and magazines that focused on crisis, in particular the *New York Times*, the *Washington Post*, *Newsweek*, and *Time*, were instrumental in arousing public concern about environmental hazards. See Marcus 1980, p. 287.

43. The Senate's ocean dumping report reads: "Most of the subjects ... are important not only environmentally but economically as well. Since much of current economic concern stems from the relative competitive position of different nations in world markets, it is important to get as many nations as possible to impose like environmental restraints upon themselves ... the U.S. domestic legislation can promote international agreement by treating the subject of ocean dumping in international waters separately. [In order to avoid law of the sea issues, the proposal for U.S. dumping regulation was based on the right to regulate transportation from U.S. ports and by U.S. registered ships]. By taking this route, the U.S. can tend to equalize out competitive position relative to European industry." (Senate Report 451, pp. 4242–4243) The issue of disadvantage to U.S economic interests was also raised by a representative from the private sector at the International Conference on Ocean Pollution. See Mark Morton, "Statement," in U.S. Senate, Subcommittee on Oceans and Atmosphere of the Committee on Commerce, International Conference on Ocean Pollution: October 18 and November 8, 1971 (92nd Congress, 2nd session), p. 27.

44. See letter from President Nixon to Sen. William V. Roth Jr., *Congressional Record: Senate*, April 1, 1971, p. 9209.

45. The reasons for preferring the OECD forum were explained on p. 57 of U.S. Senate, Committee on Foreign Relations, U.N. Conference on Human Environment—Preparations and Prospects: May 3, 4, and 5, 1972 (92nd Congress, 2nd Session).

46. Charles C. Humpstone, "Statement," in U.S. Senate, Committee on Foreign Relations, U.N. Conference on Human Environment—Preparations and Prospects: May 3, 4, and 5, 1972 (92nd Congress, 2nd Session), p. 63.

47. See Hollick 1974, p. 65. According to an inter-departmental report, U.S. high priority interests in the field of marine pollution were: preventing pollution from ship operations and from dumping, studies of ocean pollution effects, monitoring of pollutants affecting ocean quality, inventory of ocean pollution sources, economic

impact of adopting ocean pollution standards, and development of pollution control technologies. Other priorities were sustaining yields of marine living resources, long-range weather forecasting, and man-made ecological upset of the sea. See Committee on International Environmental Affairs, Task Force III, US Priority Interests in the Environmental Activities of International Organizations (1970), pp. 54–65.

48. Quoted in U.S. Senate, Committee on Foreign Relations, U.N. Conference on Human Environment—Preparations and Prospects: May 3, 4, and 5, 1972 (92nd Congress, 2nd Session), p. 68.

49. Christian A. Herter Jr., "Statement," in U.S. House of Representatives, Subcommittee on International Organizations and Movements of the Committee on Foreign Affairs, International Cooperation in the Human Environment through the United Nations: March 15 and 16, 1972 (92nd Congress, 2nd session), pp. 37–38.

50. Senate Concurrent Resolution 53 (92nd Congress, 1st session) reprinted in U.S. Senate, Committee on Foreign Relations, *U.N. Conference on Human Environment—Preparations and Prospects: May 3, 4, and 5, 1972* (92nd Congress, 2nd Session), 1972, p. 54. It was furthermore agreed that such accords should provide ways of assisting the efforts of the developing nations in meeting environmental standards and regulations. See also Springer 1988, pp. 46–52.

51. One senator had pointed out how an international agreement on ocean dumping could pave the way for domestic regulation: "Agreements reached at a 1971 international conference [i.e., the International Conference on Ocean Pollution] would be important to the 92nd Congress in its deliberations on a national policy on ocean dumping. It seems quite possible to me that such agreements as could be reached in the international arena this year would have an important bearing on actions which the Congress might need to take to bring about a desired national policy on ocean dumping." (Sen. William V. Roth Jr., *Congressional Record: Senate*, April 1, 1971, p. 9209)

52. Rep. Alton Lennon, in U.S. House of Representatives, Subcommittee on Fisheries and Wildlife Conservation and the Subcommittee on Oceanography of the Committee on Merchant Marine and Fisheries, *Ocean Dumping of Waste Materials: April 5, 6, 7, 1971* (92nd Congress, 1st session), p. 1.

Chapter 6

1. The UN's General Assembly confirmed Maurice Strong's appointment as Secretary-General of the Stockholm conference on December 7, 1970. But Strong was already in charge of the preparations at that point and had met with the Preparatory Committee.

2. From the beginning, Strong emphasized that concrete action should be taken in Stockholm. See "Opening Remarks by Maurice F. Strong, Secretary-General Designate, United Nations Conference on Human Environment, at Informal Meeting of Preparatory Committee for the Conference," United Nations Centre for Economic and Social Information CESI Note/13 (November 10, 1970).

3. For the published "basic paper" commissioned by the secretariat and produced at the United Nations Institute for Training and Research (UNITAR), see Serwer 1972, pp. 178–207.

4. For the published UNITAR report concerned with marine pollution, see Schachter and Serwer 1971.

5. Similarly, an American economist's account of one of the many environmental meetings prior to Stockholm reads: "The United States is still about five years ahead both in the dimensions of its problem and in the public awareness of what is happening. . . . The most radical and iconoclastic voices raised on behalf of the environment at the conference came from the Americans. . . . That David Brower of the Friends of the Earth should respond that way was not too surprising, but that such Establishment types as Stewart Udall, former Secretary of the Interior, or the scholarly Raymond Fosberg of the Smithsonian Institution should share his views puzzled otherwise sophisticated Europeans." (Goldman 1971, pp. 358–359) On the characteristic "doom-and-gloom" emphasis in American ecology debate and the U.S. influence on the international environmental movement, see McCormick 1989, pp. 69–87. For an overview of early American "environmental globalism" or "world-order environmentalism" literature, see the appendix to Falk 1971.

6. Francesco Di Castri, "Statement," in U.S. Senate, Committee on Commerce, and House of Representatives, Committee on Science and Astronautics, International Environmental Science: May 25 and 26, 1971 (92nd Congress, 1st session), p. 37.

7. See also Castro 1972, pp. 237–252.

8. See "Problems of Environment in India," reprinted in U.S. Senate, Committee on Commerce, and House of Representatives, Committee on Science and Astronautics, International Environmental Science: May 25 and 26, 1971 (92nd Congress, 1st session), pp. 222–229. See also Summary of the Indian National Report, in The Human Environment, Volume 2—Summaries of National Reports. Submitted in Preparation for UN Conference on the Human Environment (Woodrow Wilson International Center for Scholars, 1972), pp. 35–40. China did not prepare a national report for the Stockholm conference.

9. See "Environment and Development: The Founex Report on Development and Environment," *International Conciliation* 586 (January 1972), pp. 7–37. For a good discussion, see Juda 1979, pp. 90–107.

10. "The World View in Stockholm," *Newsweek*, June 12, 1972.

11. A columnist in Nigeria's *Lagos Daily Times* offered this view: "The idea of family planning as peddled by the Euro-American world is an attempt to keep Africa weak." (quoted in "The World View in Stockholm," *Newsweek*, June 12, 1972)

12. See CESI Note/13, UN Centre for Economic and Social Information (November 10, 1970), p. 5.

13. The so-called Oslo Convention, which regulates dumping in the North Sea, pioneered the use of black and gray lists. As in the negotiations on global ocean dumping regime, the negotiators followed the advice of GESAMP. For the black and gray lists adopted in the Oslo and Paris Conventions, see Bjerre and Hayward 1984, pp. 142–157.

14. For the study requested by Maurice Strong and presented to the preparatory committee suggesting the separation of the adjustable "technical part" from the more permanent "diplomatic part" of an international treaty, see Contini and Sand 1972.

15. For example, as Christian A. Herter, Special Assistant to the Secretary of State for Environmental Matters, explained about a then-recent attempt to reach an agreement on control of oil pollution of the oceans: "It is very difficult to persuade a number of countries that normal spillage-cleaning of tanks, this kind of thing-in the ocean produced a very serious threat in terms of the total ocean . . . a certain amount of scientific information was produced by experts pointing out the hazards to marine life of oil pollution. But it was perfectly clear that more scientific information and research on this topic was required, and the lack of it made the political process more difficult." ("Statement," in U.S. House of Representatives, Subcommittee on International Organizations and Movements of the Committee on Foreign Affairs, International Cooperation in the Human Environment through the United Nations: March 15 and 16, 1972 (92nd Congress, 2nd session), p. 45)

16. For a detailed discussion of Brazil's position at the Stockholm conference, see Guimarães 1991, pp. 147–159.

17. See, e.g., Cowan 1970.

18. This symbolic advantage of completing a dumping convention was also apparent to Hunter (1972, p. 17).

19. Algeria, Argentina, Australia, Belgium, Brazil, Britain, Canada, Chile, Cuba, Cyprus, Denmark, France, Ghana, Iceland, Iran, Italy, Japan, Madagascar, Malta, Mexico, Morocco, Netherlands, Norway, Peru, Philippines, Portugal, Soviet Union, Spain, Sweden, Turkey, United Arab Republic, United States, Yugoslavia.

20. "Report of the First Session of the Inter-Governmental Working Group on Marine Pollution, London. 14–18 June 1971," A/CONF.48/IWGMP.I/5 (June 21, 1971), p. 3.

21. "Address by the Secretary-General of the Conference—Mr. Maurice F. Strong," Annex IV to A/CONF.48/IWGMP.I/5, p. 2.

22. "Report of the Intergovernmental Working Group on Marine Pollution on Its Second Session," A/CONF.48/IWGMP.II/5 (November 22, 1971), p. 1.

23. Annex IV to A/CONF.48/IWGMP.I/5, p. 6.

24. According to Special Assistant to the Secretary of State for Environmental Affairs Christian A. Herter Jr.: "It was not felt in London that at this point it was possible to deal with what you might call coastal and estuarine pollution because there are enormously complicated problems involving jurisdictional boundaries and all sorts of things, and as you are fully aware, the so-called law of the sea conference is designed to deal with these problems of territorial jurisdiction. . . . It was felt if we can get anything done at Stockholm at all, let's keep it fairly simple." ("Statement," in U.S. Senate, International Conference on Ocean Pollution, p. 35)

25. The U.S. draft reads: "No party shall issue permits for the transportation of such material for dumping if the dumping thereof in the ocean would unreasonably degrade or endanger human health, welfare or amenities, or the marine environment,

ecological systems, or existing or future economic use of the ocean." (Article 3 (b). A/CONF.48/IWGMP.I/5, Annex 5, p. 4)

26. Principles submitted by Canada: "General Principles on Ocean Disposal. A. There should be general prohibition on use of the international areas of the oceans for the disposal of materials. B. The only materials that should be considered for disposal in such areas are those whose deleterious effects can be assessed with confidence taking into account factors such as assimilation by the food chain and distribution by physical processes such as ocean currents. General Considerations. A. There should be no assumption that the oceans have an excess assimilative capacity. B. The localization of the disposal of materials into waters not within the international area of the oceans would permit a more ready assessment of the effects. C. Materials of long persistence and toxicity should not be disposed of into any area of the sea." (A/CONF.48/IWGMP.I/5, Annex 6)

27. Algeria, Argentina, Australia, Barbados, Belgium, Brazil, Britain, Canada, Chile, Colombia, Cuba, Denmark, Ecuador, Finland, France, Gabon, Ghana, Guatemala, Iceland, India, Iran, Italy, Ivory Coast, Japan, Kenya, Libyan Arab Republic, Malaysia, Malta, Mexico, Netherlands, New Zealand, Norway, Peru, Portugal, South Africa, Soviet Union, Spain, Sweden, Tanzania, United States, Zambia.

28. Brazil also proposed that developing countries planned anti-pollution measures according to their economic possibilities, and also suggested assistance to the developing countries in controlling their marine economic resources. See A/CONF.48/IWGMP.I/5, Annex 13, pp. 1–2.

29. Article 1 reads: "The contracting parties pledge themselves to take all possible steps to prevent the pollution of the sea by substances that are liable to create hazards to human health, harm living resources and marine life, damage amenities or interfere with other legitimate uses of the sea." General criteria for issue of permits took also a firmer prohibitory and restrictive stance. Article 7 reads: "No party shall grant permits for dumping if the dumping of matter or the continued dumping thereof would [materially] endanger human health, welfare or amenities, the marine environment, living and other marine resources, ecological systems, or other legitimate uses of the sea." (A/CONF.48/IWGMP.II/5, pp. 9–10)

30. See A/CONF.48/IWGMP.II/5, Articles 2(7) and 2(8), p. 9.

31. Article 3 said: "It is forbidden to dump at sea toxic mercury, cadmium, organohalogen [and organosilicon] compounds, [and oil and derivative hydrocarbons], other than those which are rapidly converted in the sea into substances which are biologically harmless, except as noted in article IV. [The dumping of biological and chemical warfare agents and [high level] radioactive waste is also prohibited]. The dumping of other matter which has a deleterious effect on the marine environment equivalent to the properties of the matter referred to above is also prohibited." Similarly, Article 8, which referred to "special permit," although it was not yet precisely defined, stated: "For such substances as [radioactive wastes], arsenic, lead, copper and zinc, and their compounds, cyanides and fluorides, and pesticides, a special permit for each dumping shall be required." Because the question of regulatory authority over radioactive waste still was unresolved, a distinct reference to IAEA

appeared in Article 12: "Nothing in this convention supplants any recommendations designed to regulate the disposal of any material adopted by the International Atomic Energy Agency." (A/CONF.48/IWGMP.II/5, pp. 9–12)

32. Algeria, Argentina, Australia, Belgium, Britain, Canada, Denmark, Federal Republic of Germany, Finland, France, Ghana, Iceland, India, Iran, Ireland, Ivory Coast, Japan, Kenya, Malta, Mexico, Netherlands, Nigeria, Norway, Portugal, Singapore, Spain, Sweden, Tunisia, United States.

33. See article 6 in the U.S. draft (Canadian Delegation, "Composite Articles on Dumping from Vessels at Sea," April 7, 1972).

34. The parties to the Oslo Convention had agreed that the convention should not cover oil and radioactive materials. Article 14 ("The Contracting Parties pledge themselves to promote, within the competent specialized agencies and other international bodies, measures concerning the protection of the marine environment against pollution caused by oil and oily wastes, other noxious or hazardous cargoes, and radioactive materials") indicated only that the member states agreed to cooperate within the relevant forums; oil and radioactive materials accordingly were not listed in the annexes. (memos, Ministry of Foreign Affairs, Denmark, December 13, 1971, and May 1, 1972) For the text of the Oslo Convention, see Convention for the Prevention of Marine Pollution by Dumping from Ships and Aircraft, *ILM* 11 (November 1972), pp. 262–266. In 1992, the Oslo Convention merged with the so-called Paris Convention to create the Convention for the Protection of the Marine Environment of the North-East Atlantic (OSPAR).

35. See article 11 in Text of Draft Articles of a Convention for the Prevention of Marine Pollution by Dumping (IMOD/2, April 15, 1972).

36. Algeria, Australia, Belgium, Britain, Denmark, Federal Republic of Germany, France, Iceland, India, Ivory Coast, Japan, Kenya, Netherlands, Norway, Spain, Sweden, United States.

37. See "Intergovernmental Meeting on Ocean Dumping. London, 30 and 31 May, 1972," A/CONF.48/C.3/CRP.19/ (June 6, 1972), annex C (b).

38. Clearly, the ocean dumping convention was perceived to be at the top of the list of concrete accomplishments in Stockholm: "To take the positive things first: the conference resolved to establish an international convention on marine dumping." (Hawkes 1972b, p. 1308)

39. Commenting on the draft convention on ocean dumping, developing countries insisted that "the Articles failed to distinguish between developed and developing countries in terms of their relative capacity to pollute the oceans. It was feared thereby that an unfair burden would be imposed on developing countries in the event of such a convention coming into force. It was pointed out that an international law to control dumping must, in the first place, avoid authorizing present practices of dumping by industrialized countries, a possibility which has been protested by a large majority of States already." (quoted from Friedheim 1975, p. 179)

40. Eighty countries participated.

41. See also Timagenis 1980, pp. 193–195.

42. On the "double standards," see Timagenis 1980, pp. 77–79 and 178. Developing countries' concerns similarly underlie principle 23 of the Stockholm Declaration. It reads: "Without prejudice to such criteria as may be agreed upon by the international community, or to standards which will have to be determined nationally, it will be essential in all cases to consider the systems of values prevailing in each country, and the extent of the applicability of standards which are valid for the most advanced countries but which may be inappropriate and of unwarranted social cost for the developing countries." (reprinted in Rowland 1973, p. 145)

43. The technical working group produced the following text: "High-level radioactive wastes defined as unsuitable including the ecological viewpoint for dumping at sea, by the competent international body in this field." ("Report of Technical Working Party," DWS(T)7 1st Revise (3 November 1972), p. 2)

44. See DWS(T)7 1st Revise, p. 5.

45. See Convention on the Prevention of Marine Pollution by Dumping of Wastes and Other Matter, Annex I, item 6 in *ILM* 11 (November 1972), p. 1310.

46. See Convention on the Prevention of Marine Pollution by Dumping of Wastes and Other Matter, Annex II, item D in ibid., p. 1311.

47. The clause itself was a strong reason why states were willing to conclude the negotiations.

48. Because of the unsettled German issue and the refusal to the GDR of diplomatic recognition, the Soviet Union and the eastern European countries, with the exception of Rumania, were absent from the Stockholm conference.

49. House Report no. 361, Marine Protection, Research, and Sanctuaries Act of 1971 (92nd Congress, 1st session) (1971), p. 14.

Chapter 7

1. The emphasis on scientific consensus in the epistemic communities literature is also noted on p. 12 of Litfin 1994.

2. For a study of American environmental leaders and prominent science celebrities enjoying intense mass media coverage in the 1960s and the 1970s, see Goodell 1975.

3. See, e.g., Litfin 1994, p. 32.

4. For a similar critique, see Risse-Kappen 1994, p. 187.

5. It is not unusual for the building of an international regime to take 20 years or more. See Levy et al. 1995, p. 280.

6. A short decision time is likely to increase the demand for leadership in international negotiations. See Underdal 1994, p. 183.

7. For the full text of Train's statement, see Treaty Information: London Conference Agrees on Ocean Dumping Convention, Department of State Bulletin, December 18, 1972, pp. 710–711.

8. For an inductive approach to the interpretation of the national interest, see Krasner 1978, pp. 42–45.

9. The lecturers included Thor Heyerdahl, Sir Solly Zuckerman, Aurelio Peccei (cofounder of the Club of Rome), Georges Bananescu, Gunnar Myrdal, Barbara Ward, and René Dubos.

10. Note, however, that using "sticks and carrots" alone should not be considered genuine leadership: some minimum of shared values, interests, and beliefs must be present within a group of states. See, e.g., Underdal 1994, p. 179.

11. Majone (1985, p. 52) makes a similar observation in his discussion of the "pull" effects of lead countries.

12. Similarly, economists doubt that unilaterally going first and setting a good example for others to follow is a viable and effective way to reach agreement. See, e.g., Barrett 1993, pp. 459–460.

13. This hypothesis is strong in the sense that it states necessary or sufficient conditions so one single contrary case is sufficient to falsify the hypothesized relationships. See Young and Osherenko 1993a, p. ix. See also Underdal 1994, p. 192.

14. Garrett and Weingast suggest that the principle of mutual recognition (that goods and services that may legally be sold in one country should have unrestricted access to other markets within the EC) functioned as a shared belief and as such facilitated cooperation among governments.

15. On the shortcomings of interest-based analysis of leaders and entrepreneurs, see Malnes 1995.

16. No single, generally accepted, comprehensive theory of domestic politics exists today to explain international cooperation. For a rather recent overview of the dominant approaches, see Milner 1992, pp. 488–495.

17. According to the broad interpretation of the notion of egoism in interest-based regime theory, policy entrepreneurs promoting their own values and shaping public policy should be understood as egoists. For Oran Young's interpretation of self-interest, see Young 1991, pp. 293–297. On farsighted and myopic egoism, see Keohane 1984, pp. 110, 122.

18. Nelson's article "We're Making a Cesspool of the Sea," originally published in the magazine *National Wildlife*, is reprinted on pp. 30978–30979 of *Congressional Record: Senate*, September 9, 1970.

19. In the 1980s and the early 1990s, a few marine scientists opposed to the trend toward all-out protection of the oceans occasionally tried to reach out to public opinion through the *New York Times* and other publications with a wide circulation. (See Charles Osterberg, "Seas: To Waste or Not," *New York Times*, August 9, 1981; Michael A. Champ, "The Ocean and Waste Disposal," *The World and I* 5, April 1990.) The need for marine scientists to participate more intensely in policy relevant dialogue about marine pollution issues with the public and decision makers is stressed in Walsh 1982.

20. One illustration of how the "the dying oceans" idea simplified a much more complex reality is Sen. Hollings's opening statement to the second session of the International Conference on Ocean Pollution: "As we meet, the destruction of the oceans continues. Not only does it continue-it accelerates . . . [experts] know full well that the Mediterranean has a death sentence hanging over it. That the Atlantic,

the Pacific, the Black Sea, the Red Sea, and the Baltic are all in danger of destruction." (U.S. Senate, International Conference on Ocean Pollution, p. 39) And, a second illustration: "The shores of United States waterways have been so invaded by pollutants, that evidence indicates the nation already has a number of "dead" seas: some parts of Lake Erie, the Houston Ship Canal, San Pedro Bay, the upper Delaware River, off the Hudson Estuary, and others. Elsewhere on earth, the Caspian Sea, the Baltic Sea, the Irish Sea, the Sounds between Sweden and Denmark, the North Sea, and the Sea of Japan can be found in varying stages of entrophication. Perhaps the most shocking example of the developed nations' apparent lack of concern about the quality of their waters is the Cuyahoga River in Cleveland, Ohio, which actually burst into flame in 1969." (Weinstein-Bacal 1978, p. 875.)

21. See, e.g., "New York's Dead Sea," *Newsweek*, February 23, 1970, p. 86.

22. See Grayson and Shepard 1973, p. 66.

23. "Danish delegates said the Baltic Sea is already subject to such oxygen starvation as to be almost dead, and that the North Sea is rapidly becoming the same." This quote from Friendly 1971 appears on p. 199 of U.S. Senate, Ocean Waste Disposal.

24. For example, Heyerdahl concluded his speech at the International Institute for Environmental Affairs' seminar series held concurrently with the Stockholm Conference with these words: ". . . the revolving ocean, indispensable and yet vulnerable, will forever remain a common human heritage" (Heyerdahl 1973, p. 63). In an essay titled "Stop Killing Our Oceans," Sen. Gaylord Nelson wrote: "Finally, all nations together must establish an International Policy on the Sea that sacrifices narrow self-interest for the protection of this vast domain that is a common heritage of all mankind." (*Congressional Record: Senate*, February 8, 1971, p. 2036)

25. See Weart 1988, p. 325.

26. See Linsky 1990, p. 206.

27. See Kingdon 1984, pp. 62–63.

28. According to a close observer, "it was not until the [the Preparatory Committee] met for the third time in September of 1971 that it had become clear that the job of educating the public and their political leaders had already been partly accomplished through the explosion of public interest in environmental matters, and that the Stockholm conference would have to produce some concrete action if it were to be taken seriously by observers around the world" (Rowland 1973, p. 87).

29. On the success of the Stockholm conference, see, e.g., Hawkes 1972b, pp. 1308–1310; "Mr Strong's Recipe," *Nature* 237 (June 23, 1972), pp. 418–419. One analyst concluded that the preparations for and subsequent convening of the Stockholm conference had "apparently heightened environmental consciousness internationally" (Soroos 1981, p. 30).

30. "Only One Earth" was the conference's epigram.

31. Many representatives of the scientific community were skeptical about the quality of scientific knowledge about environmental impact of pollutants. See, e.g., Myrdal 1973, pp. 70–71. Strong's position marked a compromise between environmental crusaders and their critics. For a discussion, see Handelman et al. 1974, p. 76.

Chapter 8

1. For the situation of those four nations at the time, see table 2.3.

2. Pro-dumping and anti-dumping governments had later been unable to reach an agreement at a 1966 meeting held in Lisbon. See van Weers et al. 1982, p. 451.

3. These shipments and the subsequent storage of plutonium in Japan caused (and still cause) considerable domestic and international controversy. See O'Neill 1999.

4. "Mariana Islanders Protest Plans By Japan to Dump Atomic Waste," *New York Times*, August 3, 1980; Van Dyke et al. 1984, p. 742; Branch 1984, pp. 329–330.

5. Plans by the United States will be discussed below.

6. See also "Japan: Seabed Dumping Delayed as Other Nations Object," *Nuclear News* 24, March 1981, 61–62.

7. Japan's desire to act in a responsible way internationally motivated this and later decisions on the issue (interview, Takao Kuramochi, August 30, 1991). This consideration might have been given some weight by Japanese governments as the United States had urged Japan to join, or even provide international leadership behind, efforts to protect the environment; see, e.g., Friedman 1991. In 1980, however, Japanese fishermen, fearing adverse effects from dumping on the fish stocks that constituted their livelihood, threatened to use their boats to hinder dumping; see Trumbull 1980; see also Junkerman 1981, p. 32. At the same time, the government of the Commonwealth of the Northern Marianas reportedly warned that Japanese vessels would be excluded from its fishing zone should dumping operations be commenced; see Finn 1983b, p. 216. The following considerations have also been suggested to explain Japan's decision not to dump: "Japan has been cultivating an image of peace and conciliation in an effort to live down its prewar reputation as the Pacific Basin's strong-arm bully; [and] it has made substantial investments in the island nations and territories, and wants to make more." (Carter 1987, p. 364)

8. In 1980, officials of the Japanese Science and Technology Agency noted it was still "not necessary to get approval of foreign governments" (Kirk 1980, p. 4). In 1981, a leading official of the same agency reportedly (Kamm 1981) said: "We will continue to have a plan to dump. We did not give up the plan."

9. See also Bakalian 1984.

10. See also Shabecoff 1982.

11. In 1982, the Department of Energy estimated that land disposal of 60,000 metric tons of slightly radioactive soil would cost well over $100 million, but ocean disposal less than $10 million. See Roberts 1982, p. 774. In 1984, it was estimated that dumping nine retired submarines would cost about $5.2 million, whereas burying them on land would cost $13.3 million (Trupp 1984, p. 34). Both of the Navy's cost estimates (which did not include the costs of monitoring over a period of several hundred years, and which presented scientific documentation claiming that no harm would be inflicted on the marine environment and human health) were criticized by U.S. groups opposing the Navy's plans. See "Joint Comments of Environmental and Other Citizen Organizations in Response to the Department of Navy's

Draft Environmental Impact Statement on the Disposal of Decommissioned, Defueled Naval Submarine Reactor Plants" (submitted to the Department of the Navy by the Center for Law and Social Policy and the Oceanic Society, June 30, 1983).

12. See Roberts 1982.

13. See Carter 1980, pp. 1495–1997; Norman 1982, pp. 1217–1219. A report sponsored by the EPA concluded that "the disposal of radioactive waste into the ocean evokes strong feelings. There are those who feel that radioactive materials should be completely prohibited from deep ocean disposal." It was further assessed, however, that "it is naive to believe that all other countries will accept a position of not permitting the disposal of packaged low-level waste when for certain countries it is the only option available to them." (Hagen 1980), pp. 7-1 and 7-2.

14. "Contamination Survey Set for Boston Harbor," *New York Times*, September 18, 1982.

15. For the expectations of the EPA researchers, see "Offshore Waste Study Begun," *New York Times*, September 21, 1982. Compared to samples taken in areas where no dumping had occurred, the dumpings were not found to have caused detectable levels of radioactivity in Massachusetts Bay. (See "Radwaste in Massachusetts Bay," *BioScience* 33 (February 1983), p. 87.) For doubts about the scientific value of these measurements, see Sielen 1988, p. 26.

16. U.S. House of Representatives, Committee on Merchant Marine and Fisheries, Disposal of Decommissioned Nuclear Submarines: October 19, 1982 (97th Congress, 2nd session); Subcommittee of the Committee on Government Operations, Ocean Dumping of Radioactive Waste off the Pacific Coast: October 7, 1980 (96th Congress, 2nd session).

17. The Oceanic Society represented 26 environmental and public-interest groups (including Friends of the Earth, Greenpeace USA, the National Audubon Society, the Sierra Club, and the Union of Concerned Scientists) opposed to the Department of Navy's proposal. The Oceanic Society and other groups also organized a series of citizen workshops on the proposal to dump aged nuclear submarines. Workshops were to be held in Boston, in Washington, in Winston-Salem and Beaufort, North Caroline, in Charleston, South Carolina, in Eureka, California, and in Seattle. See "Joint Comments of Environmental and Other Citizen Organizations in Response to the Department of Navy's Draft Environmental Impact Statement on the Disposal of Decommissioned, Defueled Naval Submarine Reactor Plants"; see also "Oceanic Society Leads Opposition to Nuclear Dumping," *Oceans* 16 (September-October 1983), p. 70.

18. See also Joseph A. Davis, "Legislation to Strengthen Rules on Ocean Dumping Approved by the House," *Congressional Quarterly, Weekly Report* 40 (September 25, 1982), p. 2386.

19. At the end of the moratorium, the bill required EPA to make a comprehensive environmental statement before a permit for ocean disposal of low-level radioactive waste could be issued. Congress was given 30 days to review and block issuance of each permit which made it very unlikely that any dumping permit would be issued

within a short period of time. Failure by Congress to act within 30 days on a permit request results in a legislative veto, resulting in a denial of the permit. The bill's sponsors deliberately chose this procedure because they expected it was extremely unlikely that Congress within so brief time could consider a permit.

20. See "Navy Prefers to Bury Subs," *Science News* 125 (May 26–June 9, 1984), p. 358. See also Trupp 1984, pp. 34–35.

21. On the Seabed Working Group of the NEA (OECD), see Curtis 1985 and Deese 1977.

22. Greenpeace has since its beginning been an anti-nuclear organization working to reduce nuclear weapons, nuclear weapons testing, and nuclear power. Its nuclear-free seas campaign has aimed at "ridding the seas of nuclear weapons" (GREENPEACE . . . for a cleaner, safer earth). A newsletter released in relation to Greenpeace's campaign against radwaste disposal explained that "a ban on dumping of radioactive waste would hasten the death of the nuclear industry, particularly in the U.K." (quoted on p. 79 of Rippon 1983).

23. A major international scandal followed when in 1985 the *Rainbow Warrior* was blown up by French agents in Auckland Harbor.

24. On Greenpeace's campaign within Britain, see Blowers et al. 1991, pp. 74–85. A good source of information is the Greenpeace documentary film *Desperate Measures*, which documents the Greenpeace campaign from its beginning in 1978 through 1982.

25. No one was injured in the accident. See "Dutch Ship Stops Dumping Nuclear Waste," *New York Times*, August 30, 1982.

26. See "A Dutch Ship Resumes Dumping Nuclear Waste Off Northern Spain," *New York Times*, August 31, 1982.

27. Quoted on p. 427 of Wassermann 1985b.

28. "Dutch to Stop Dumping Nuclear Wastes at Sea," *New York Times*, September 23, 1982.

29. See van Weers et al. 1982, pp. 461–462.

30. For a short history of the "greening" of the Netherlands, see Bennett 1991.

31. See Article 15 (2).

32. The London Convention originally puts the onus of proof on those wanting to halt pollution. See Article 1.

33. Other Pacific Basin nations, such as Fiji, had chosen not to do so because they have considered the regime too lenient as demonstrated by Japan's claim that its proposed dumping was in accord with the convention. Kiribati (the former Gilbert Islands, which gained independence from Britain in 1979) became a member of the London Convention on June 11, 1982; Nauru (a former United Nations trust territory that became independent in 1968) became a member on August 25, 1982.

34. For the discussion, see LDC 1983a, pp. 19–30.

35. See LDC 1982.

36. Hollister criticized the misinterpretation of his scientific work on the geological effects of deep sea currents by Davis et al. Hollister was one of the chief American spokespersons of sub-seabed disposal of high-level radioactive wastes. For an example of Hollister's many publications on this waste disposal option, see Hinga et al. 1982. On Hollister's involvement in this issue, see Miles et al. 1985, pp. 21–23.

37. See NACOA 1984.

38. See also "London Dumping Convention—7th Consultative Meeting," *Environmental Policy and Law* 10 (1983): 83–85.

39. None of the few developing countries producing radioactive waste conducted ocean dumping.

40. See also Short 1984, pp. 14–18.

41. In 1984 a French cargo vessel carrying nuclear waste sank in the English Channel.

42. On the incident in the English Channel, see Wassermann 1985a, pp. 178–181.

43. See "Four Unions Back Ban on A-Waste Dumping," *The Guardian* (Manchester), April 7, 1983. See also Samstag 1983.

44. Many described the ship as "Greenpeace-proof." See, e.g., Samstag 1983.

45. See Wright 1983a and Debelius 1983.

46. See Samstag 1985.

47. See "Boycott of Nuclear Dumping at Sea," *Times* (London), September 10, 1983; "Slater Fears Nuclear Waste May be Dumped on Seabed in Submarine," *The Guardian*, September 10, 1983.

48. Blowers et al. 1991, p. 82.

49. A minor controversy among the four transport unions determined to stop the dumping and the TUC took place prior to the positions of the transport unions and the TUC were made public. See Edwards 1985.

50. See LDC 1984, pp. 17–33.

51. See Curtis 1984, pp. 68–69.

52. See "London Dumping Convention—7th Consultative Meeting," *Environmental Policy and Law* 10 (1983), pp. 83–85.

53. Invited experts should cover the following disciplines: radiological protection, radiation biology, radioecology, radioactive waste management, modeling, marine biology, physical oceanography, marine geochemistry, marine ecology, marine geology.

54. "Introduction of Report Prepared by the Panel of Experts. The Disposal of Low-Level Radioactive Waste at Sea. (Review of Scientific and Technical Considerations)," LDC/PRAD.1/2/Add.2 (May 1, 1985), p. 17.

55. Upper bound signifies the maximum amount of total human irradiation permitted from a certain source. On the question of risk to deep sea fauna, the report stated: "The results obtained so far indicate that there is no risk of significant damage to local populations of deep sea fauna at or near the North-East Atlantic dump site." (LDC. 1985d, p. 137)

56. Since 1981 Greenpeace International had status as observer at consultative meetings of the regime. The organization was allowed to make oral statements and submit written material. See LDC 1981, pp. 3–4.

57. (LDC. 1985a, p. 26)

58. See LDC 1985c, pp. 16–41.

59. A few months later, Canada changed its negative vote to a positive vote.

60. According to the meeting report (members of the global ocean dumping regime adopt the meeting report at the end of their meeting, and it might therefore not be an entirely accurate account of what was said), Britain told the meeting that "such tactics had brought the Ninth Consultative Meeting very close to the point at which some Contracting Parties might have to reconsider the terms of their participation in the Convention." For the British and American statements, see LDC 1985c, annex 5, pp. 10–11.

61. The chairman, Geoffrey Holland of Canada, later circulated a letter to all members in which he pointed out that "the integrity of the Convention could suffer if a sequence of amendments were adopted that were unacceptable to some Contracting Parties." He also noted that decisions on the annexes should be based on scientific and technical considerations. See Miles 1987, p. 49.

62. See Spiller and Hayden 1988, p. 352.

63. For an environmental NGO's view on the significance of this decision, see Curtis 1986, p. 14.

64. See LDC 1986, pp. 22–33. For statements made by contracting parties during the discussion on radwaste disposal, see Annex 10 of that report.

65. According to an announcement made on May 23, 1989 in the British parliament (quoted on p. 324 of Birnie and Boyle): "The Government have decided not to resume sea-disposal of drummed radioactive waste, including waste of military origin. None the less, the Government intend to keep open this option for large items arising from decommissioning operations, although they have taken no decisions about how redundant nuclear submarines will be disposed of."

66. See Fairhall 1989 and Pienaar 1989.

67. In 1989, the EPA was redrafting regulations for ocean disposal of low-level radioactive wastes. See Galpin et al. 1989, pp. 177–180.

68. On concern in Massachusetts, see Dumanoski and McLaughlin 1991. On concern about Pacific dumpings, see "Atomic Waste Reported Leaking in Ocean Sanctuary Off California," *New York Times*, May 7, 1990); "US Sees Threat in Nuclear Dump," *Boston Globe*, May 6, 1990; "Radioactive Waste Threatens Sanctuary," *Washington Post*, May 6, 1990; Bishop 1991.

69. I attended the meeting as an observer.

70. See Stairs and Taylor 1992, pp. 123–127.

71. On the weight attributed to the concept of assimilative capacity in Britain, see Boehmer-Christiansen 1990, pp. 139–149.

72. See Sielen 1988.

73. See LC/IGPRAD 1993, Annex 2, p. 12.

74. For an analysis of Russia's implementation of its LC commitments, see Stokke 1998, pp. 475–517.

75. In addition, it seems likely, as U.S. and Danish officials and Greenpeace representatives pointed out in interviews, that Japan did not want to be out of step with the rest of the international community on this issue.

76. On the Russian position at the meeting, see Vartanov and Hollister 1997.

77. Source: Schoon 1994. But the British government was unconvinced about the risk and unattractiveness of radwaste disposal. Hence, the British minister said: "The scientific evidence shows that dumping at sea, carried out under controlled conditions, causes no harm to the marine environment and poses no threat to human health. This has been confirmed by careful monitoring over many years and studies have shown it to be the best practicable environmental option for the disposal of certain types of radioactive waste."

78. "La France signe la convention de Londres sur l'interdiction de l'immersion des déchets nucléaires," *Le Monde*, December 22, 1993.

Chapter 9

1. On two-level games, see Putnam 1988.

2. I thank Edward L. Miles for stressing this point.

3. In their UNITAR report concerned with marine pollution, Schachter and Serwer (1971, p. 110) wrote: "Action in the area of marine dumping need not, however, come only through the initiative of international organizations and governments. In a number of countries, action on pollution problems of the "dangerous practices" type has been stimulated largely through the initiatives of private citizens and concerned organizations who have taken pollution problems to court. The effectiveness of such private actions varies with the situation, but they must be considered an important mode of action where governments which are responsible for controlling pollution are participants in practices which may cause pollution. An international mechanism for handling complaints and grievances from private groups as well as governments might contribute to the control not only of marine dumping of wastes but to the control of other dangerous practices as well. Moreover, such a mechanism might be one form in which problems of international concern could be adequately discussed from both the technical and legal points of view."

4. For the rules, see Article 15 (2).

5. In the case of drift net fishing, for example, the United States has backed pressure with threats of banning import of Japanese marine products, perhaps including pearls, a major import product (Weisman 1991).

6. For instance, according to Robert A. Nisbet, cooperation is "joint or collaborative behavior that is directed toward some goal and in which there is common interest or hope of reward" (*International Encyclopedia of the Social Sciences* (Library of Congress, 1968), volume 3, p. 385).

7. According to Keohane's formal definition (1984, pp. 51–52), "intergovernmental cooperation takes place when the policies actually followed by one government are regarded by its partners as facilitating realization of their own objectives, as the result of a process of policy coordination." For a discussion of this definition, see Milner 1992, pp. 467–470.

8. As I mentioned in chapter 3, realist and neoliberal studies of power used to achieve a common good focus on malign and benign hegemons. For a brief discussion of power in the service of the common good, see Krasner 1983, pp. 13–14.

9. Keohane (1984, p. 53) disagrees with this.

10. I put "radiation-free" in quotation marks, since several radionuclides (e.g. potassium-40 and uranium) exist naturally in the oceans.

11. See P. Haas 1989, 1992b. But note the conclusion of an epistemic-community-oriented case study in international regulation of commercial whaling (Peterson 1992, p. 182): "Yet the epistemic community of conservation-minded cetologists only briefly enjoyed predominant influence over policy. Most of the time, the influence of cetologists was outweighed by that of other groups, the industry managers until the mid-1960s and the environmentalists after the mid-1970s."

12. GESAMP (1992) stressed the need for a holistic and multi-sectoral framework (i.e., land, air, and water) for the management of radioactive wastes, that wastes were unavoidably produced despite clean technologies and recycling, and that radwaste disposal had not presented "appreciable risks" to humans or the environment.

13. On marine scientists' discussion of the concept of assimilative capacity, see Stebbing 1992.

14. A long and heated debate on the concept(s) of the precautionary principle took place in the journal *Marine Pollution Bulletin* after John Gray, a leading member of GESAMP wrote that "the precautionary principle is entirely an administrative and legislative matter and has nothing to do with science" (Gray 1990, p. 174). See letters in *Marine Pollution Bulletin* 21 (December 1990) by Paul Johnston and Mark Simmonds (p. 402), Alf B. Josefson (p. 598), John Lawrence and D. Taylor (pp. 598–599), and John S. Gray (pp. 599–600); see also Earll 1992; Peterman and M'Gonigle 1992. Other scientists have also criticized the precautionary principle; see, e.g., Milne 1993, pp. 34–37. More generally, see O'Riordan and Cameron 1994.

15. M. D. Hill, Head of Assessments Department, U.K. National Radiological Protection Board, quoted in House of Lords, Select Committee on the European Communities, 19th Report: Radioactive Waste Management. Session 1987-88. Minutes of Evidence (1988), p. 34.

16. On the historic development of public concern about the marine environment, see Waldichuk 1982, pp. 37–75.

17. For an overview and a discussion of this literature, see Breitmeier and Rittberger 2000.

18. But Princen and his colleagues have also stated, in the case of the Antarctic, that "to promote regime change . . . the environmental NGOs supplanted the scientific NGOs" (Princen et al. 1995, p. 48).

19. "The Greening of British Politics," *The Economist*, March 3, 1990.
20. See, e.g., Young 1989, pp. 371–372.
21. See McCormick 1989 and Boxer 1982.
22. On the insurance industry and global warming, see Flavin 1994.
23. See Oye and Maxwell 1995, pp. 191–221.
24. For other examples, see Raustiala 1997.
25. See, e.g., Peterson 1992.
26. From 1985 to 1990, membership in Greenpeace increased from 1.4 million to 6.75 million worldwide and annual revenues went from $24 million to some $100 million (Princen and Finger 1994, p. 2). In 1990, it employed approximately a thousand people in 23 offices located on four continents, including Asia and Latin America, and owned seven vessels. See Greenpeace International 1990, p. 9. Greenpeace evidently is a global organization and so are, among others, the World Wide Fund for Nature (WWF) and the Friends of the Earth. In fact, ENGOs sometimes have more issue-specific resources at their disposal than states, international organizations, and regime secretariats, and ENGO delegations are often among the largest attending environmental regime meetings.

Chapter 10

1. Those who believe that U.S. hegemony has declined expect that material power resources are likely to play a less important role when future regimes will be built. See, e.g., Keohane 1984.
2. For a general discussion, see Rhodes 1985. For an examination of the role of self-interest and the role of social norms in achieving social order, see Elster 1989.
3. See Cox 1983.
4. Others similarly note that Western norms are most likely to become internationalized. See, e.g., Finnemore and Sikkink 1998, pp. 906–907.
5. See, e.g., Lundqvist 1980.
6. On the scant attention paid to costs of EC water policy in the early 1970s, see Richardson 1994.
7. This is well illustrated by the international efforts to regulate CFCs. See Oye and Maxwell 1995, pp. 191–221.
8. Risk perception analysts and psychologists have repeatedly confirmed the general public's fear of nuclear waste and the prevalence of not-in-my-backyard (NIMBY) sentiments. See, e.g., Slovic et al. 1991 and Carter 1987. France is the exception.
9. Similarly, as two advocates of subseabed disposal of high-level radioactive wastes recently observed (Hollister and Nadis 1998, p. 44), public resentment makes this option "a tough sell."
10. For examples and discussion, see Gamson and Modigliani 1989.

11. See Hasenclever et al. 1997, pp. 211–224.

12. See Underdal 1995.

13. For the U.S., see Skolnikoff 1997. For the EU, see Forum Inputs to the European Union Climate Change Policy Strategy, produced by the European Consultative Forum on the Environment and Sustainable Development, an advisory forum of the European Commission (http://europa.eu.int/comm/dg11/forum/report0399.pdf).

14. On the role of ideas on development strategies in developing countries, see Sikkink 1991. On the possible role of ideas in the EU, see Risse-Kappen 1996.

15. On ideas and macro-economic policy, see Hall 1989b, pp. 361–391. On the role of ideas on foreign policy, see Risse-Kappen 1994. On the impact of international norms on human rights in Germany, see Checkel 1999. On how domestic structures determine the accessibility of states to transnational actors and their ultimate policy impact, see Risse-Kappen 1995b, pp. 288–293.

16. On alternative models, see Underdal 1998.

17. For a critical review of the literature, see Moravcsik 1999.

18. According to Keohane (1984, p. 73), for instance, regimes "do not necessarily improve world welfare" and "are not ipso facto 'good.'" For an example from international trade, see p. 177 of Haus 1991.

19. For a recent controversial development in which an environmental coalition has been a major driving force, see Burke et al. 1994.

20. The issue of scientific uncertainty did not loom large in the early 1970s because a belief that science could provide irrefutable, objective answers to questions about environmental damage prevailed at that time. The discovery of PCBs, for instance, raised expectations that measurement techniques existed that were able to accurately detect minute concentrations of pollutants. See Jensen 1972. UN officials involved in the preparations for the Stockholm Conference hoped that international environment protection would "be a highly technical matter" (Contini and Sand 1972, p. 56).

21. This protocol has not yet entered into force.

22. Although one of the LC's preambles refers to the assimilative capacity notion, this does not imply that scientific knowledge about this capacity should guide policy making. If future cooperation should have a scientific foundation at all, it was just a matter of course that the convention assumed the existence of such an assimilative capacity (interview, Ole Vagn Olsen, March 19, 1992). Other concerns also played a role. Governments that worried about the economic costs of pollution control and lack of sufficient disposal alternatives supported that the convention made a reference to the concept of assimilative capacity. The U.S. Department of Commerce, for example, strongly supported that the convention recognized the existence of the assimilative capacity of the ocean (interview, Robert J. McManus, August 29, 1991).

23. ACMP was set up by the International Council for the Exploration of the Sea. For ACMP's approach, see Report of the ICES Advisory Committee on Marine Pollution (Cooperative Research Report 167, ICES, Copenhagen, 1989), pp. 124–145.

24. The LC's Annex 3 does list such considerations; however, as McManus has noted (1983, p. 124), it "is totally lacking in prescriptive content."

25. For an early argument stressing that knowledge of the marine environment is too limited for opening up the oceans to more waste disposal, see Kamlet 1981. For Greenpeace's criticism of the assimilative capacity concept, see "Critical Review of GESAMP Report No. 45 on "Global Strategies for Marine Environmental Protection," submitted by Greenpeace International to Fourteenth Consultative Meeting of the London Dumping Convention (LDC14/Inf.29, IMO, London, 1991).

26. Divergence between the view of experts and that of the public and policy makers is by no means restricted to the issue of radwaste disposal. For other examples, see *Dædalus* 119 (fall 1990).

27. For instance, a survey from the mid 1980s asking the British public who they would trust to supervise nuclear waste (Campbell and Forbes 1985, p. 5) found that "MPs [Members of Parliament] of any party, managers from the nuclear industry, and anyone in the government, police, or armed forces were "the last people" that interviewees would trust. Public trust was limited to groups like Greenpeace, "independent scientists," "women," and "investigative journalists." There was "deep public scepticism about the feasibility of properly monitoring (any radioactive waste disposal) system. Most people felt that bribery, corruption or hit squads might be employed to shut up dissenters."

28. Thus, most contributions on epistemic communities "seem to believe an idea is utterly neutral when championed by an epistemic community" (Jacobsen 1995, p. 288).

References

Allen-Mills, Tony. 1979. "Atomic Waste Dumpers Foil Saboteurs." *Daily Telegraph* (London), July 12.

Andresen, Steinar. 1989. "Science and Politics in the International Management of Whales." *Marine Policy* 13 (April): 99–117.

Arbose, Jules. 1972. "91 Nations Agree on Convention to Control Dumping in Oceans." *New York Times*, November 14.

Ardill, John. 1983. "Unions to Block Dumping of Nuclear Waste in Atlantic." *Guardian* (Manchester), June 18.

Aron, Raymond. 1966. *Peace and War: A Theory of International Relations*. Doubleday.

Axelrod, Robert, and Robert O. Keohane. 1986. "Achieving Cooperation under Anarchy: Strategies and Institutions." In *Cooperation under Anarchy*, ed. K. Oye. Princeton University Press.

Bakalian, Allan. 1984. "Regulation and Control of United States Dumping: A Decade of Progress, An Appraisal for the Future." *Harvard Environmental Law Review* 8: 193–256.

Baldwin, David A., ed. 1993. *Neorealism and Neoliberalism: The Contemporary Debate*. Columbia University Press.

Barrett, Scott. 1993. "International Cooperation for Environmental Protection." In *Economics of the Environment*, ed. R. Dorfman and N. Dorfman. Norton.

Barston, R. P., and P.W. Birnie. 1980. "The Marine Environment." In *The Maritime Dimension*, ed. R. Barston and P. Birnie. Allen & Unwin.

Bedlow, Robert. 1972. "57 Nations Sign Pollution Treaty to Control Dumping at Sea."*Daily Telegraph* (London), November 11.

Bennett, Graham. 1991. "The History of the Dutch National Environmental Policy Plan." *Environment* 33 (September): 6–9 and 31–33.

Bewers, J. M. 1995. "The Declining Influence of Science on Marine Environmental Policy." *Chemistry and Ecology* 10: 9–23.

Bewers, J. M., and C. J. R. Garrett. 1987. "Analysis of the Issues Related to Sea Dumping of Radioactive Wastes." *Marine Policy* 11 (April): 105–124.

Birnie, Patricia W., and Alan E. Boyle. 1992. *International Law and the Environment*. Clarendon.

Bishop, Katherine. 1991. "U.S. to Determine if Radioactive Waste in Pacific Presents Danger." *New York Times*, January 20.

Bjerre, F., and P. A. Hayward. 1984. "The Role and Activities of the Oslo and Paris Commissions." In *Environmental Protection*, ed. T. Lack. Ellis Horwood.

Blowers, Andrew, David Lowry, and Barry D. Solomon. 1991. *The International Politics of Nuclear Waste*. Macmillan.

Boehmer-Christiansen, Sonja. 1990. "Environmental Quality Objectives versus Uniform Emission Standards." In *The North Sea*, ed. D. Freestone and T. Ijlstra. Graham & Trotman/Nijhoff.

Böhme, Eckart. 1972. "The Use of the Seabed as a Dumping Site—Viewed from the Outcome of the FAO Technical Conference on Marine Pollution, Rome 1970." In *From the Law of the Sea towards an Ocean Space Regime*, ed. E. Böhme and M. Kehden. Werkhefte der Forschungsstelle für Völkerrecht und ausländisches öffentliches Recht der Universität Hamburg.

Bourke, Gerald. 1983. "Europeans Seek Answers to Nuclear Waste Buildup." *Chemical Engineering* 90 (February 7): 25–26.

Boxer, Baruch. 1982. "Mediterranean Pollution: Problem and Response." *Ocean Development and International Law Journal* 10: 315–356.

Branch, James B. 1984. "The Waste Bin: Nuclear Waste Dumping and Storage in the Pacific." *AMBIO* 13: 327–333.

Breitmeier, Helmut, and Volker Rittberger. 2000. "Environmental NGOs in an Emerging Global Civil Society." In *The Global Environment in the Twenty-First Century: Prospects for International Cooperation*, ed. P. Chasek. United Nations University Press.

Brown, Paul. 1985a. "Britain Seeks Allies to Lift Nuclear Dumping Truce." *Guardian* (Manchester), September 21, 1985.

Brown, Paul. 1985b. "Open-Ended Nuclear Dumping Ban: Britain Loses Strong Rearguard Action as Vote Switches Burden of Proof." *Guardian* (Manchester), September 27.

Brown, Paul. 1985c. "UK Threatens to Withdraw from Convention on Nuclear Dumping." *Guardian* (Manchester), September 25.

Burke, William T., Mark Freeberg, and Edward L. Miles. 1994. "United Nations Resolutions on Driftnet Fishing: An Unsustainable Precedent for High Seas and Coastal Fisheries Management." *Ocean Development and International Law* 25: 127–186.

Burley, Anne-Marie. 1993. "Regulating the World: Multilateralism, International Law, and the Projection of the New Deal Regulatory State." In *Multilateralism Matters*, ed. J. Ruggie. Columbia University Press.

Caldwell, Lynton K. 1970. "Government and Environmental Quality." In *No Deposit—No Return*, ed. H. Johnson. Addison-Wesley.

Caldwell, Lynton K. 1984. *International Environmental Policy: Emergence and Dimensions*. Duke University Press.

Callick, Rowan. 1980. "A Storm Beneath the Calm." *Far Eastern Economic Review* (November 7): 40–41.

Campbell, Duncan, and Patrick Forbes. 1985. "£100 Million To Be Made As Nuclear Waste Dumpers Scramble To Get Rich Quick." *New Statesman*, October 18.

Camplin, W. C., and H. D. Hill. 1986. "Sea Dumping of Solid Radioactive Waste: A New Assessment." *Radioactive Waste Management and the Nuclear Fuel Cycle* 7 (August): 233–251.

Carr, E. H. 1939. *The Twenty Years' Crisis: An Introduction to the Study of International Relations*. Harper and Row.

Carter, Luther J. 1980. "Navy Considers Scuttling Old Nuclear Subs." *Science* 209 (September 26): 1495–1497.

Carter, Luther J. 1987. *Nuclear Imperatives and Public Trust: Dealing with Radioactive Waste*. Resources for the Future.

Castaing, Michel. 1993. "L'immersion de déchets radioactifs est définitivement interdite." *Le Monde*, November 16.

Castro, João Augusto de Araujo. 1972. "Environment and Development: The Case of the Developing Countries." In *World Eco-Crisis*, ed. D. Kay and E. Skolnikoff. University of Wisconsin Press.

Cemlyn-Jones, Bill. 1983. "Protest at Nuclear Waste Plan." *Guardian* (Manchester), July 13.

CEQ (Council on Environmental Quality). 1970. *Ocean Dumping—A National Policy*.

Checkel, Jeffrey T. 1997. *Ideas and International Political Change: Soviet/Russian Behavior and the End of the Cold War*. Yale University Press.

Checkel, Jeffrey T. 1999. "Norms, Institutions, and National Identity in Contemporary Europe." *International Studies Quarterly* 43 (March): 83–114.

Christenson, Reo M. 1976. *Challenge and Decision: Political Issues of Our Time*. Harper and Row.

Clark, R. B. 1989a. "Ocean Dumping." *Marine Pollution Bulletin* 20 (June): 295.

Clark, R. B. 1989b. "The Mediterranean, the Media, and the Public Interest." *Marine Pollution Bulletin* 20 (August): 369–372.

Commoner, Barry. 1970. "The Ecological Facts of Life." In *The Ecological Conscience*, ed. R. Disch. Spectrum Books.

Commoner, Barry. 1971. *The Closing Circle: Nature, Man, and Technology*. Knopf.

Conlan, Timothy, David Beam, and Margaret Wrightson. 1995. "Policy Models and Political Change: Insights from the Passage of Tax Reform." In *The New Politics of Public Policy*, ed. M. Landy and M. Levin. John Hopkins University Press.

Contini, Paolo, and Peter H. Sand. 1972. "Methods to Expedite Environment Protection: International Ecostandards." *American Journal of International Law* 66: 37–59.

Cooper, Richard N. 1989. "International Cooperation in Public Health as a Prologue to Macroeconomic Cooperation." In *Can Nations Agree?* ed. R. Cooper. Brookings Institution.

Cousteau, Jacques. 1971. "Our Oceans Are Dying." *New York Times*, November 14.

Cowan, Edward. 1970. "U.S., Canada Asked to Save the Lakes: 8 States Join 3 Provinces in Antipollution Drive." *New York Times*, September 11.

Cox, Robert W. 1983. "Gramsci, Hegemony and International Relations: An Essay in Method." *Millennium* 12 (summer): 162–175.

Cruickshank, Andrew. 1983. Dumping in Deep Water?" *Nuclear Engineering International* 28 (September): 13–14.

Curtis, Clifton E. 1983a. "Ocean Dumping Nations Vote Radwaste Suspension." *Oceanus* 26 (spring): 76–78.

Curtis, Clifton E. 1983b. "Radwaste Dumping Delayed: An International Moratorium Keeps Nuclear Wastes at Bay." *Oceans* 16 (May-June): 22–23.

Curtis, Clifton E. 1984. "Radwaste Disposal Risks Assessed at LDC Meeting." *Oceanus* 27 (summer): 68–71.

Curtis, Clifton E. 1985. "Legality of Seabed Disposal of High-Level Radioactive Wastes under the London Dumping Convention." *Ocean Development and International Law* 14: 383–415.

Curtis, Clifton E. 1986. "Radioactive Wastes: Reflections on International Policy Developments Under the London Dumping Convention." Paper presented at Sixth International Ocean Disposal Symposium, Asilomar Conference Center, Pacific Grove, California.

de Almeida, Miguel A. Ozorio. 1972. "The Confrontation between Problems of Development and Environment." *International Conciliation* 586 (January): 37–56.

Debelius, Harry. 1983. "Spaniards Pelt British Embassy." *Times* (London), July 12.

Deere-Jones, Tim. 1991. "Back to the Land: The Sea-to-Land Transfer of Radioactive Pollution." *Ecologist* 21 (January-February): 18–23.

Deese, David A. 1977. "Seabed Emplacement and Political Reality." *Oceanus* 20 (winter): 47–63.

Deese, David A. 1978. *Nuclear Power and Radioactive Waste: A Sub-Seabed Disposal Option?* Lexington Books.

Deutsch, Karl W., Sidney A. Burrell, Robert A. Kann, Maurice Lee Jr., Martin Lichterman, Raymond E. Lindgren, Francis L. Loewenheim, and Richard W. Van Wagenen. 1957. *Political Community and the North Atlantic Area*. Princeton University Press.

Dibblin, Jane. 1985a. "Paddling in the Nuclear Pool." *New Statesman*, March 1.

Dibblin, Jane. 1985b. "Britain Is in the Dock." *New Statesman*, September 20.

Disch, Robert, ed. 1970. *The Ecological Conscience: Values for Survival*. Spectrum Books.

Dobrynin, Anatoly. 1995. *In Confidence: Moscow's Ambassador to America's Six Cold War Presidents*. Time Books.

Douglas, Mary, and Aaron Wildavsky. 1982. *Risk and Culture: An Essay on the Selection of Technological and Environmental Dangers*. University of California Press.

Duchêne, François. 1994. *Jean Monnet: The First Statesman of Interdependence*. Norton.

Dumanoski, Dianne, and Jeff McLaughlin. 1991. "Probe of Ocean Waste Site Urged." *Boston Globe*, August 17.

Duncan, Rodney N. 1974. "The 1972 Convention on the Prevention of Marine Pollution by Dumping of Wastes at Sea." *Journal of Maritime Law and Commerce* 5 (January): 299–315.

Dyer, Robert S. 1981. "Sea Disposal of Nuclear Waste: A Brief History." In *Nuclear Waste Management: The Ocean Alternative*, ed. T. Jackson. Pergamon.

Earll, R. C. 1992. "Commonsense and the Precautionary Principle." *Marine Pollution Bulletin* 24 (April): 182–186.

Eckstein, Harry. 1975. "Case Study and Theory in Political Science." In *Handbook of Political Science—Volume 7: Strategies of Inquiry*, ed. F. Greenstein and N. Polsby. Addison-Wesley.

Edwards, Rob. 1983. "Wasting the Ocean." *New Statesman*, July 1.

Edwards, Rob. 1985. "TUC Muffles Union Discord over Sea-Dumping of Nuclear Waste." *New Statesman*, September 6.

Efron, Edith. 1984. *The Apocalyptics: How Environmental Politics Control What We Know about Cancer*. Touchstone.

Ehrlich, Paul. 1972. "Eco-Catastrophe!" In *Society and Environment*, ed. R. Campbell and J. Wade. Allyn and Bacon.

Elster, Jon. 1983. *Explaining Technical Change: A Case Study in the Philosophy of Science*. Cambridge University Press.

Elster, Jon. 1989. *The Cement of Society: A Study of Social Order*. Cambridge University Press.

Enloe, Cynthia H. 1975. *The Politics of Pollution in a Comparative Perspective: Ecology and Power in Four Nations*. Longman.

Fairhall, David. 1989. "MoD Favours Scuttling Old Nuclear Subs." *Guardian* (Manchester), April 13.

Falk, Richard A. 1971. *This Endangered Planet*. Random House.

Farvar, M. Taghi, and John P. Milton, eds. 1972. *The Careless Technology. Ecology and International Development*. Natural History Press.

Finn, Daniel P. 1983a. "International Law and Scientific Consultation on Radioactive Waste Disposal in the Ocean." In *Wastes in the Ocean. Volume 3: Radioactive Wastes and the Ocean*, ed. P. Kilho Park et al. Wiley.

Finn, Daniel P. 1983b. "Nuclear Waste Management Activities in the Pacific Basin and Regional Cooperation on the Nuclear Fuel Cycle." *Ocean Development and International Law Journal* 13: 213–246.

Finnemore, Martha, and Kathryn Sikkink. 1998. "International Norm Dynamics and Political Change." *International Organization* 52 (autumn): 887–917.

Flateboe, Connie. 1970. "The UNESCO Manifesto." In *Ecotactics*. Pocket Books.

Flavin, Christopher. "Storm Warnings: Climate Change Hits the Insurance Industry." *World Watch* 7 (1994): 10–20.

Flowers, R. H. 1989. "Radioactive Waste Management in the United Kingdom." In *Proceedings of the 1989 Joint International Waste Management Conference—Volume 1: Law and Intermediate Level Radioactive Waste Management*, ed. S. Slate et al. American Society of Mechanical Engineering.

Fortner, Robert S. 1994. *Public Diplomacy and International Politics*. Praeger.

Franck, Thomas M. 1990. *The Power of Legitimacy among Nations*. Oxford University Press.

Friedheim, Robert L. 1975. "Ocean Ecology and the World Political System." In *Who Protects the Ocean?* ed. J. Hargrove. West.

Friedman, Thoman L. 1991. "Baker to Japan: Share the Global Burden." *International Herald Tribune*, November 12.

Friendly, Alfred. 1971. "U.S. Plan Fails at NATO Conference—Ban on Ocean Dumping Is Rejected." *Washington Post*, April 20.

Frye, Alton. 1962. *The Hazards of Atomic Wastes: Perspectives and Proposals on Oceanic Disposal*. Public Affairs Press.

Galpin, F. L., W. F. Holcomb, J. M. Gruhlke, and D. J. Egan Jr. 1989. "The U.S. Environmental Protection Agency's Radioactive Waste Disposal Regulatory Activities." In *Proceedings of the 1989 Joint International Waste Management Conference—Volume 2: High Level Radioactive Waste and Spent Fuel Management*, ed. S. Slate et al. American Society of Mechanical Engineering.

Gamson, William A., and Andre Modigliani. 1989. "Media Discourse and Public Opinion on Nuclear Power: A Constructionist Approach." *American Journal of Sociology* 95 (July): 1–37.

Gardner, Richard N. 1972. "The Role of the UN in Environmental Problems." *International Organization* 26 (spring): 237–259.

Garrett, Geoffrey, and Barry R. Weingast. 1993. "Ideas, Interests, and Institutions: Constructing the European Community's Internal Market." In *Ideas and Foreign Policy*, ed. J. Goldstein and R. Keohane. Cornell University Press.

George, Alexander L., and Timothy J. McKeown. 1985. "Case Studies and Theories of Organizational Decision Making." In *Advances in Information Processing in Organizations. Volume 2*, ed. R. Coulam and R. Smith. JAI.

GESAMP (Group of Experts on the Scientific Aspects on Marine Pollution). 1990. *The State of the Marine Environment*. Blackwell.

GESAMP. 1991. Global Strategies for Marine Environmental Protection. GESAMP Reports and Studies no. 45, IMO.

GESAMP. 1992. Can There Be a Common Framework for Managing Radioactive and Non-Radioactive Substances to Protect the Marine Environment? Addendum 1, GESAMP Reports and Studies no. 45, IMO.

Gillette, Robert. 1972. "Politics of the Ocean: View from the Inside." *Science* 178 (November 17): 729–732 and 793–795.

Gilpin, Robert. 1981. *War and Change in World Politics*. Cambridge University Press.

Goble, R. L. 1983. "Time Scales and the Problem of Radioactive Waste." In *Equity Issues in Radioactive Waste Management*, ed. R. Kasperson. Oelgeschlager, Gunn and Hain.

Goldman, Marshall I. 1971. "Has the Environment a Future?" *Nation*, October 18.

Goldstein, Judith. 1993. *Ideas, Interests, and American Trade Policy*. Cornell University Press.

Goodell, Rae. 1975. *The Visible Scientists*. Little, Brown.

Goth, Louis A. 1974. "The Great Lakes Are Scarcely Great, But Getting Better." *New York Times*, June 9.

Gray, Bernard. 1996. "Nuclear Submarines Left to Rust." *Financial Times*, June 13.

Gray, John. 1990. "Statistics and the Precautionary Principle." *Marine Pollution Bulletin* 21 (April): 174–176.

Grayson, Melvin J., and Thomas R. Shepard Jr. 1973. *The Disaster Lobby: Prophets of Ecological Doom and Other Absurdities*. Follett.

Green, Harold P., and L. Marc Zell. 1982. "Federal-State Conflict in Nuclear Waste Management: The Legal Bases." In *The Politics of Nuclear Waste*, ed. E. Colglazier Jr. Pergamon.

Greenpeace International. 1990. Political and Social Impact of a Resumption of Radioactive Waste Dumping at Sea. Statement submitted to Third Meeting of IGPRAD.

Grieco, Joseph M. 1990. *Cooperation among Nations: Europe, America, and Non-Tariff Barriers to Trade*. Cornell University Press.

Guimarães, Roberto P. 1991. *The Ecopolitics of Development in the Third World: Politics and Environment in Brazil*. Lunne Reinner.

Gusfield, Joseph R. 1981. *The Culture of Public Problems: Drinking-Driving and the Symbolic Order*. University of Chicago Press.

Haas, Ernst B. 1958. *The Uniting of Europe*. Stanford University Press.

Haas, Ernst B. 1980. "Why Collaborate? Issue-Linkage and International Regimes." *World Politics* 32 (April): 357–405.

Haas, Ernst B. 1983. "Words Can Hurt You: Or, Who Said What to Whom about Regimes." In *International Regimes*, ed. S. Krasner. Cornell University Press.

Haas, Ernst B. 1990. *When Knowledge Is Power: Three Models of Change in International Organizations*. University of California Press.

Haas, Ernst B., Mary Pat Williams, and Don Babai. 1977. *Scientists and World Order: The Use of Technical Information in International Organizations*. University of California Press.

Haas, Peter M. 1989. "Do Regimes Matter? Epistemic Communities and Mediterranean Pollution Control." *International Organization* 43 (summer): 377–403.

Haas, Peter M. 1990a. *Saving the Mediterranean: The Politics of International Environmental Cooperation*. Columbia University Press.

Haas, Peter M. 1990b. "Obtaining International Environmental Protection through Epistemic Consensus." *Millennium* 19 (winter): 347–363.

Haas, Peter M. 1992a. "Introduction: Epistemic Communities and International Policy Coordination." *International Organization* 46 (winter): 1–35.

Haas, Peter M. 1992b. "Banning Chlorofluorocarbons: Epistemic Community Efforts to Protect Stratospheric Ozone." *International Organization* 46 (winter): 187–224.

Hagen, Amelia A. 1980. An Analysis of International Issues Associated with Ocean Disposal of Low-Level Radioactive Waste. MITRE Corp.

Hagen, Amelia A. 1983. "History of Low-Level Radioactive Waste Disposal in the Sea." In *Wastes in the Ocean. Volume 3: Radioactive Wastes and the Ocean*, ed. P. Kilho Park et al. Wiley.

Haggard, Stephan, and Beth A. Simmons. 1987. "Theories of International Regimes." *International Organization* 41 (summer): 491–517.

Hall, Peter A., ed. 1989a. *The Political Power of Economic Ideas: Keynesianism across Countries*. Princeton University Press.

Hall, Peter A. 1989b. "Conclusion: The Politics of Keynesian Ideas." In *The Political Power of Economic Ideas*, ed. P. Hall. Princeton University Press.

Hamilton, E. I. 1986. "Science—A Time of Change?" *Marine Pollution Bulletin* 17 (July): 295–298.

Handelman, John R., Howard B. Shapiro, and John A. Vasquez. 1974. *Introductory Case Studies for International Relations: Vietnam; The Middle East; the Environmental Crisis*. Rand-McNally.

Hansen, Anders. 1993. "Greenpeace and Press Coverage of Environmental Issues." In *The Mass Media and Environmental Issues*, ed. A. Hansen. Leicester University Press.

Harwood, Michael. 1971. "We Are Killing the Sea around Us." *New York Times Magazine*, October 24.

Hasenclever, Andreas, Peter Mayer, and Volker Rittberger. 1997. *Theories of International Regimes*. Cambridge University Press.

Hasler, Arthur D., and Bruce Ingersoll. 1968. "Dwindling Lakes." Reprinted in *Eco-Crisis*, ed. C. Johnson (Wiley, 1970).

Haus, Leah. 1991. "The East European Countries and GATT: The Role of Realism, Mercantilism, and Regime Theory in Explaining East-West Trade Negotiations." *International Organization* 45 (spring): 163–182.

Hawkes, Nigel. 1972a. "Human Environment Conference: Search for a Modus Vivendi." *Science* 175 (February 18): 736–738.

Hawkes, Nigel. 1972b. "Stockholm: Politicking, Confusion, but Some Agreements Reached." *Science* 176 (June 23): 1308–1310.

Hawkes, Nigel. 1972c. "Sea Dumping Talks Clear Snags." *Observer* (Manchester), November 12.

Heclo, Hugh. 1978. "Issue Networks and the Executive Establishment." In *The New American Political System*, ed. A. King. American Enterprise Institute.

Heyerdahl, Thor. 1973. "How Vulnerable Is the Ocean?" In *Who Speaks for Earth?* ed. M. Strong. Norton.

Hiatt, Fred. 1993. "After Yeltsin Visit, Russia Is Dumping A-Waste off Japan." *International Herald Tribune*, October 18.

Hill, Gladwin. 1972. "China Denounces U.S. on Pollution: Reparations to Poor Nations Demanded in Stockholm." *New York Times*, June 11.

Hinga, K.R., G. Ross Heath, D. Richard Anderson, and Charles D. Hollister. 1982. "Disposal of High-Level Radioactive Wastes by Burial in the Sea Floor: This Method May Be Technically and Environmentally Feasible." *Environmental Science and Technology* 16 (January): 27A–37A.

Hirschman, Albert. 1984. "Policymaking and Policy Analysis in Latin America—A Return Journey." In *Essays in Trespassing*, ed. A. Hirschman. Cambridge University Press.

Hirschman, Albert O. 1989. "How the Keynesian Revolution Was Exported from the United States, and Other Comments." In *The Political Power of Economic Ideas*, ed. P. Hall. Princeton University Press.

Holcomb, W. F. 1982. "A History of Ocean Disposal of Packaged Low-Level Radioactive Waste." *Nuclear Safety* 23 (March-April): 183–197.

Hollick, Ann L. 1974. "Bureaucrats at Sea." In *New Era of Ocean Politics*, ed. A. Hollick and R. Osgood. Johns Hopkins University Press.

Hollister, Charles D., and Steven Nadis. 1998. "Burial of Radioactive Waste under the Seabed." *Scientific American* 278 (January): 40–45.

House of Lords, Select Committee on the European Communities. 1988. 19th Report: Radioactive Waste Management. Session 1987–88. Minutes of Evidence.

Hunter, Lawson A. W. 1972. "Prospects for an Ocean Dumping Convention." In *The Question of An Ocean Dumping Convention* (no. 2, Studies in Transnational Legal Policy, American Society of International Law).

IAEA (International Atomic Energy Agency). 1993. Risk Comparisons Relevant to the Sea Disposal of Low Level Radioactive Waste.

IAEA. 1994. Nuclear Power, Nuclear Fuel Cycle and Waste Management: Status and Trends 1994.

IMO (International Maritime Organization). 2000. Summary of Status of Conventions as of February 29, 2000. http://www.imo.org/imo/convent/ summary.htm.

Ishihara, Takehiko. 1982. "Ocean Dumping of Low-Level Wastes in Japan: Past and Future." In proceedings of Symposium on Waste Management, Tucson.

Jacobsen, John K. 1995. "Much Ado about Ideas: The Cognitive Factor in Economic Policy." *World Politics* 47 (January): 283–310.

Jensen, Sören. 1972. "The PCB Story." *AMBIO* 1 (August): 123–131.

Jervis, R. 1976. *Perception and Misperception in International Politics.* Princeton University Press.

Joesten, Joachim. 1969. *Wem gehört der Ozean? Politiker, Wirtschaftler und moderne Piraten greifen nach den Weltmeeren.* Südwest Verlag.

Johnson, H., ed. 1970. *No Deposit—No Return.* Addison-Wesley.

Johnston, Paul, and Mark Simmonds. 1990. Letter. *Marine Pollution Bulletin* 21 (1990): 402.

Jones, Peter. 1989. "Plans to Dump Nuclear Subs at Sea." *Marine Pollution Bulletin* 20 (June): 251.

Jönsson, Christer. 1993. "Cognitive Factors in Explaining Regime Dynamics." In *Regime Theory and International Relations*, ed. V. Rittberger. Clarendon.

Josefson, Alf B. 1990. Letter. *Marine Pollution Bulletin* 21 (December): 598.

Juda, Lawrence. 1979. "International Environmental Concern: Perspectives of and Implications for Developing States." In *The Global Predicament*, ed. D. Orr and M. Soroos. University of North Carolina Press.

Junkerman, John. 1981. "Deep-Sixing the Atom." *The Progressive* 45 (December): 32.

Kamlet, Kenneth S. 1981. "The Oceans as Waste Space: The Rebuttal." *Oceanus* 24 (spring): 10–17.

Kamm, Henry. 1981. "Islanders Fight Japan's Plan to Dump Atom Waste." *New York Times*, March 18.

Katzenstein, Peter J., Robert O. Keohane, and Stephen D. Krasner. 1998. "*International Organization* and the Study of World Politics." *International Organization* 52 (autumn): 645–685.

Kaul, Inge, Isabelle Grunberg, and Marc A. Stern, eds. 1999. *Global Public Goods: International Cooperation in the 21st Century.* Oxford University Press.

Kegley, Charles W. Jr., ed. 1995. *Controversies in International Relations Theory: Realism and the Neorealism Challenge.* St. Martin's Press.

Kelman, Steven. 1990. "Why Public Ideas Matter." In *The Power of Public Ideas*, ed. R. Reich. Harvard University Press.

Keohane, Robert O. 1980. "The Theory of Hegemonic Stability and Changes in International Economic Regimes, 1967–1977." In *Change in the International System*, ed. O. Holsti et al. Westview.

Keohane, Robert O. 1984. *After Hegemony: Cooperation and Discord in the World Political Economy.* Princeton University Press.

Keohane, Robert O. 1988. "International Institutions: Two Approaches." *International Studies Quarterly* 32 (December): 379–396.

Keohane, Robert O. 1997. "What Can Political Scientists Contribute to an Understanding of Environmental Policy?" *Degrees of Change* (newsletter from Global Change Integrated Assessment Program, Department of Engineering and Public Policy, Carnegie Mellon University) 9 (May): 1–4.

Keohane, Robert O., and Joseph S. Nye. 1977. *Power and Interdependence: World Politics in Transition*. Little, Brown.

King, Gary, Robert O. Keohane, and Sidney Verba. 1994. *Designing Social Inquiry: Scientific Inference in Qualitative Research*. Princeton University Press.

Kingdon, John W. 1984. *Agendas, Alternatives, and Public Policies*. Little, Brown.

Kirk, Don. 1980. "Double Standards in Japan's Nuclear Policy." *New Statesman*, September 5.

Kirton, Allan. 1977. "Developing Country View of Environmental Issues." In *Law of the Sea*, ed. E. Miles and J. Gamble Jr. Ballinger.

Knauss, John A. 1973. "Ocean Pollution: Status and Prognostication." In *Law of the Sea*, ed. J. Gamble Jr. and G. Pontecorvo. Ballinger.

Krasner, Stephen D. 1978. *Defending the National Interest: Raw Materials Investments and U.S. Foreign Policy*. Princeton University Press.

Krasner, Stephen D. 1983. "Structural Causes and Regime Consequences: Regimes as Intervening Variables." In *International Regimes*, ed. S. Krasner. Cornell University Press.

Krasner, Stephen D. 1985. *Structural Conflict: The Third World against Global Liberalism*. University of California Press.

Krasner, Stephen D. 1991. "Global Communications and National Power: Life on the Pareto Frontier." *World Politics* 43 (April): 336–365.

Kratochwil, Friedrich V. 1989. *Rules, Norms, and Decisions: On the Conditions of Practical and Legal Reasoning in International Relations and Domestic Affairs*. Cambridge University Press.

Kremenyuk, Victor A. 1991. "The Emerging System of International Negotiation." In *International Negotiation*, ed. V. Kremenyuk. Jossey-Bass.

Lapp, Ralph E. 1979. *The Radiation Controversy*. Reddy Communications.

Laws, David. 1990. "The Antarctic Minerals Regime Negotiations." In *Nine Case Studies in International Environmental Negotiation*, ed. L. Susskind et al. Program on Negotiation, Harvard Law School.

LC (London Convention). 1993. Report of the Sixteenth Consultative Meeting. Report 16/14, December 15.

LC. 1997. Compilation of the Full Texts of the London Convention 1972 and of the 1996 Protocol Thereto. LC.2/Circ.380, March 12.

LC/IGPRAD. 1993. Report of the Sixth Meeting of the Inter-Governmental Panel of Experts on Radioactive Waste Disposal at Sea. 6/5 (August 31).

LDC (London Dumping Convention). 1978. Report of the Third Consultative Meeting. LDC III/12, October 24.

LDC. 1981. Report of the Sixth Consultative Meeting. LDC VI/12 (November 10).

LDC. 1982. Evaluation of Oceanic Radioactive Dumping Programmes. Submitted jointly by Kiribati and Nauru. LDC7/INF.2 (September 23).

LDC. 1983a. Report of the Seventh Consultative Meeting. LDC 7/12 (March 9).

LDC. 1983b. Critical Studies and Comments to the Report "Evaluation of Oceanic Radioactive Dumping Programmes." Submitted by France. LDC 8/5 (December 9).

LDC. 1984. Report of the Eighth Consultative Meeting. LDC 8/10 (March 8).

LDC. 1985a. Report of Intersessional Activities Relating to the Disposal of Radioactive Wastes at Sea, Including the Final Report of the Scientific Review Report of the Expanded Panel Meeting. LDC 9/4 (June 24).

LDC. 1985b. The Provisions of the London Dumping Convention, 1972 and Decisions Made by the Consultative Meetings of the Contracting Parties. LDC9/INF.2 (May 28).

LDC. 1985c. Report of the Ninth Consultative Meeting. LDC 9/12 (October 18).

LDC. 1985d. Introduction of Report Prepared by the Panel of Experts. LDC/PRAD.1/2 (April 12).

LDC 1985e. Introduction of Report Prepared by the Panel of Experts. The Disposal of Low-Level Radioactive Waste at Sea. (Review of Scientific and Technical Considerations). LDC/PRAD.1/2/Add.2 (May 1).

LDC. 1986. Report of the Tenth Consultative Meeting. LDC 10/15 (November 5).

LDC. 1991. Report of the 14th Consultative Meeting. LDC 14/16 (December 30).

Leitzell, Terry L. 1973. "The Ocean Dumping Convention—A Hopeful Beginning." *San Diego Law Review* 10: 502–512.

Lettow, Charles F. 1974. "The Control of Marine Pollution." In *Federal Environmental Law*, ed. E. Dolgin and T. Guilbert. West.

Levy, Marc A., Oran R. Young, and Michael Zürn. 1995. "The Study of International Regimes." *European Journal of International Relations* 1 (September): 267–330.

Lewis, Anthony. 1972. "One Confused Earth." *New York Times*, June 17.

Liberatore, Angela. 1999. *The Management of Uncertainty: Learning from Chernobyl*. Gordon and Breach.

Liefferink, Duncan. 1996. *Environment and the Nation State: The Netherlands, the European Union and Acid Rain*. Manchester University Press.

Linsky, Martin. 1990. "The Media and Public Deliberation." In *The Power of Public Ideas*, ed. R. Reich. Harvard University Press.

Lipschutz, Ronnie. 1992. "Restructuring World Politics: The Emergence of Global Civil Society." *Millennium* 21: 389–420.

Litfin, Karen T. 1994. *Ozone Discourses: Science and Politics in Global Environmental Cooperation*. Columbia University Press.

Lumsdaine, Joseph A. 1976. "Ocean Dumping Regulation: An Overview." *Ecology Law Quarterly* 5: 753–792.

Lundqvist, Lennart J. 1980. *The Hare and the Tortoise: Clean Air Policies in the United States and Sweden*. University of Michigan Press.

Maddox, John. 1972. *The Doomsday Syndrome*. McGraw-Hill.

Magnuson, Warren G. 1971. "Remarks." In *International Environmental Science: May 25 and 26, 1971* (U.S. Senate, Committee on Commerce, and House of Representatives, Committee on Science and Astronautics, 92nd Congress, 1st session).

Majone, Giandomenico. 1975. "On the Notion of Political Feasibility." *European Journal of Political Research* 3: 259–274.

Majone, Giandomenico. 1985. "The International Dimension." In *Regulating Industrial Risks*, ed. H. Otway and M. Peltu. Butterworths.

Majone, Giandomenico. 1989. *Evidence, Argument and Persuasion in the Policy Process*. Yale University Press.

Majone, Giandomenico. 1996a. *Regulating Europe*. Routledge.

Majone, Giandomenico. 1996b. "Public Policy and Administration: Ideas, Interests and Institutions." In *A New Handbook of Political Science*, ed. R. Goodin. Oxford University Press.

Malnes, Raino. 1995. "'Leader' and 'Entrepreneur' in International Negotiations: A Conceptual Analysis." *European Journal of International Relations* 1 (March): 87–112.

Marcus, Alfred. 1980. "Environmental Protection Agency." In *The Politics of Regulation*, ed. J. Wilson. Basic Books.

Marks, Gary, Liesbet Hooghe, and Kermit Blank. 1996. "European Integration from the 1980s: State-Centric *v.* Multi-level Governance." *Journal of Common Market Studies* 34 (September): 341–378.

Mathismoen, Ole. 1994. "Stortinget Roser Bellona." *Aftenposten*, June 11.

Mazuzan, George T., and J. Samuel Walker. 1985. *Controlling the Atom: The Beginnings of Nuclear Regulation 1946–1962*. University of California Press.

McCormick, John. 1989. *Reclaiming Paradise: The Global Environmental Movement*. Indiana University Press.

McCraw, Thomas K. 1984. *Prophets of Regulation*. Harvard University Press.

McDougal, Myres S., and William T. Burke. 1985. *The Public Order of the Oceans: A Contemporary International Law of the Sea*. New Haven Press. Originally published by Yale University Press in 1962.

McManus, Robert J. 1973. "The New Law on Ocean Dumping. Statute and Treaty." *Oceanus* 6 (September-October): 26–32.

McManus, Robert J. 1982. "Legal Aspects of Land-Based Sources of Marine Pollution." In *The New Nationalism and the Use of Common Spaces*, ed. J. Charney. Allanheld, Osmun.

McManus, Robert J. 1983. "Ocean Dumping: Standards in Action." In *Environmental Protection*, ed. D. Kay and H. Jacobson. Allanheld, Osmun.

Mendelsohn, Allan I. 1972. "Ocean Pollution and the 1972 United Nations Conference on the Environment." *Journal of Maritime Law and Commerce* 3 (January): 385–398.

M'Gonigle, R. Michael, and Mark W. Zacher. 1979. *Pollution, Politics, and International Law: Tankers at Sea*. University of California Press.

Miles, Edward L. 1987. *Science, Politics, and International Ocean Management: The Uses of Scientific Knowledge in International Negotiations.* University of California.

Miles, Edward L., Kai N. Lee, and Elaine M. Carlin. 1985. *Nuclear Waste Disposal under the Seabed: Assessing the Policy Issues.* University of California.

Miller, Gary J. 1997. "The Impact of Economics on Contemporary Political Science." *Journal of Economic Literature* 35 (September): 1173–1204.

Miller, H. Crane. 1973. "Ocean Dumping—Prelude and Fugue." *Journal of Maritime Law and Commerce* 5 (October): 51–75.

Milne, Alex. 1993. "The Perils of Green Pessimism." *New Scientist* 12 (June): 34–37.

Milner, Helen. 1992. "International Theories of Cooperation among Nations: Strengths and Weaknesses." *World Politics* 44 (April): 466–496.

Mitchell, Ronald B. 1994. "Regime Design Matters: Intentional Oil Pollution and Treaty Compliance." *International Organization* 48 (summer): 425–458.

Mitrany, David. 1966. *A Working Peace System.* Quadrangle.

Moore, Mark H. 1990. "What Sort of Ideas Become Public Ideas?" In *The Power of Public Ideas,* ed. R. Reich. Harvard University Press.

Moore, Mark H. 1995. *Creating Public Value: Strategic Management in Government.* Harvard University Press.

Moravcsik, Andrew. 1999. "A New Statecraft? Supranational Entrepreneurs and International Cooperation." *International Organization* 52 (spring): 267–306.

Morgenthau, Hans J. 1948. *Politics among Nations: The Struggle for Power and Peace.* Knopf.

Morris, Michael. 1978. "'Dangerous' Waste Dumped." *Guardian* (Manchester), July 25.

Moynihan, Daniel Patrick. 1978. "The United States in Opposition." In *The First World and The Third World,* ed. K. Brunner. University of Rochester.

Mueller, John. 1993. "The Impact of Ideas on Grand Strategy." In *The Domestic Bases of Grand Strategy,* ed. R. Rosecrance and A. Stein. Cornell University Press.

Myrdal, Gunnar. 1973. "Economics of an Improved Environment." In *Who Speaks for Earth?* ed. M. Strong. Norton.

NACOA (National Advisory Committee on Oceans and Atmosphere). 1984. Nuclear Waste Management and the Use of the Sea.

Nadelmann, Ethan A. 1990. "Global Prohibition Regimes: The Evolution of Norms in International Society." *International Organization* 44 (autumn): 479–526.

Nauke, Manfred, and Geoffrey L. Holland. 1992. "The Role and Development of Global Marine Conventions: Two Case Histories." *Marine Pollution Bulletin* 25: 74–79.

Nelkin, Dorothy, and Michael Pollak. 1981. *The Atom Besieged: Extraparliamentary Dissent in France and Germany.* MIT Press.

Newman, Barry. 1973. "The Sea: Pollution of Oceans Is Enormous Threat, But Few People Care." *Wall Street Journal*, October 2.

Nobile, Philip, and John Deedy, eds. 1972. *The Complete Ecology Fact Book*. Anchor Books.

Norman, Colin. 1982. "U.S. Considers Ocean Dumping of Radwastes." *Science* 215 (March 5): 1217–1219.

Norris, Robert S., and William M. Arkin. 1998. "Known Nuclear Tests Worldwide, 1945–1998." *Bulletin of the Atomic Scientists* 54 (November-December): 1–7.

Norton, M. G. 1981. "The Oslo and London Dumping Conventions." *Marine Pollution Bulletin* 12 (May): 145–149.

Nye, Joseph S. Jr. 1988. "Neorealism and Neoliberalism." *World Politics* 40 (January): 235–251.

Odell, John S. 1982. *U.S. International Monetary Policy: Markets, Power and Ideas as Sources of Change*. Princeton University Press.

Olson, Mancur. 1965. *The Logic of Collective Action: Public Goods and the Theory of Groups*. Harvard University Press.

O'Neill, Kate. 1999. "International Nuclear Waste Transportation: Flashpoints, Controversies, and Lessons." *Environment* 41 (May): 12–15 and 34–39.

Onuf, Nicholas G. 1989. *World of Our Making: Rules and Rule in Social Theory and International Relations*. University of South Carolina Press.

O'Riordan, Timothy, and James Cameron, eds. 1994. *Interpreting the Precautionary Principle*. Earthscan.

Orren, Gary R. 1990. "Beyond Self-Interest." In *The Power of Public Ideas*, ed. R. Reich. Harvard University Press.

Osherenko, Gail, and Oran R. Young. 1993. "The Formation of International Regimes: Hypotheses and Cases." In *Polar Politics*, ed. O. Young and G. Osherenko. Cornell University Press.

Osterberg, Charles. 1981. "Seas: To Waste or Not." *New York Times*, August 9.

Ostrom, Elinor. 1990. *Governing the Commons: The Evolution of Institutions for Collective Action*. Cambridge University Press.

Oye, Kenneth A. 1986. "Explaining Cooperation under Anarchy: Hypothesis and Strategies." In *Cooperation under Anarchy*, ed. K. Oye. Princeton University Press.

Oye, Kenneth A., and James H. Maxwell. 1995. "Self-Interest and Environmental Management." In *Local Commons and Global Interdependence*, ed. R. Keohane and E. Ostrom. Sage.

Parker, Frank L. 1988. "Low-Level Radioactive Waste Disposal." In *Low-Level Radioactive Waste Regulation*, ed. M. Burns. Lewis.

Pearce, Fred. 1983. "Seamen Pull the Plug on Radioactive Dumping." *New Scientist*, June 30.

Pearce, Fred. 1991. *Green Warriors: The People and the Politics Behind the Environmental Revolution*. Bodley Head.

Pearson, Charles S. 1975a. "Environmental Policy and the Ocean." In proceedings of Perspectives on Ocean Policy—Conference on Conflict and Order in Ocean Relations, Airlie, Virginia.

Pearson, Charles S. 1975b. *International Marine Environment Policy: The Economic Dimension*. John Hopkins University Press.

Peterman, Randall M., and Michael M'Gonigle. 1992. "Statistical Power Analysis and the Precautionary Principle." *Marine Pollution Bulletin* 24 (May): 231–234.

Peterson, M. J. 1992. "Whaling, Cetologists, Environmentalists, and the International Management of Whaling." *International Organization* 46 (winter): 147–186.

Pienaar, John. 1989. "Nuclear Subs May Be Scuttled." *Independent* (Dublin), April 13.

Pitt, David E. 1993a. "Pentagon Fights Wider Ocean-Dumping Ban." *New York Times*, September 26.

Pitt, David E. 1993b. "U.S. to Press for Ban on Nuclear Dumping at Sea." *New York Times*, November 2.

Polsby, Nelson W. 1984. *Political Innovation in America: The Politics of Policy Initiation*. Yale University Press.

Porges, Ralph. 1971. Statement. In United States Senate, *Ocean Waste Disposal: March 2, 3; April 15, 21, 22, and 28, 1971* (92nd Congress, 1st Session).

Powell, Robert. 1991. "Absolute and Relative Gains in International Relations Theory." *American Political Science Review* 85 (December): 1303–1320.

Power, Paul F. 1979. "The Carter Anti-Plutonium Policy." *Energy Policy* 7 (September): 215–231.

Powers, Charles F., and Andrew Robertson. 1966. "The Aging Great Lakes." Reprinted in *Man and the Ecosphere* (Freeman, 1971).

Pravdic, Velimir. 1981. *GESAMP, The First Dozen Years*. UNEP.

Preston, A. 1983. "Deep-Sea Disposal of Radioactive Wastes." In *Wastes in the Ocean. Volume 3: Radioactive Wastes and the Ocean*, ed. P. Kilho Park et al. Wiley.

Price, Richard. 1995. "A Genealogy of the Chemical Weapons Taboo." *International Organization* 49 (winter): 73–103.

Price, Richard. 1998. "Reversing the Gun Sights: Transnational Civil Society Targets Land Mines." *International Organization* 53 (summer): 613–644.

Princen, Thomas. 1994. "NGOs: Creating A Niche in Environmental Diplomacy." In *Environmental NGOs in World Politics*, ed. T. Princen and M. Finger. Routledge.

Princen, Thomas, and Matthias Finger, eds. 1994. *Environmental NGOs in World Politics*. Routledge.

Princen, Thomas, Matthias Finger, and Jack P. Manno. 1994. "Translational Linkages." In *Environmental NGOs in World Politics*, ed. T. Princen and M. Finger. Routledge.

Princen, Thomas, Matthias Finger, and Jack Manno. 1995. "Nongovernmental Organizations in World Environmental Politics." *International Environmental Affairs* 7 (winter): 42–58.

Pryor, Taylor A. 1970. "The Sea." In *No Deposit—No Return*, ed. H. Johnson. Addison-Wesley.

Putnam, Robert D. 1988. "Diplomacy and Domestic Politics: The Logic of Two-Level Games." *International Organization* 42 (summer): 427–460.

Ramey, James T. 1971. Statement. In U.S. House of Representatives, Subcommittee on Fisheries and Wildlife Conservation and the Subcommittee on Oceanography of the Committee on Merchant Marine and Fisheries, *Ocean Dumping of Waste Materials: April 5, 6, 7, 1971* (92nd Congress, 1st session).

Raustiala, Kal. 1997. "States, NGOs, and International Environmental Institutions." *International Studies Quarterly* 41 (December): 719–740.

Reich, Robert B., ed. 1990. *The Power of Public Ideas*. Harvard University Press.

Rhodes, Steven E. 1985. *The Economist's View of the World: Government, Markets, and Public Policy*. Cambridge University Press.

Richardson, Jeremy. 1994. "EU Water Policy: Uncertain Agendas, Shifting Networks and Complex Coalitions." *Environmental Politics* 3 (winter): 139–167.

Ringius, Lasse. 1996. "The Environmental Action Plan Approach: A Milestone in Pollution Control on the Baltic Sea." In *Baltic Environmental Cooperation*, ed. R. Hjorth. Department of Water and Environmental Studies, Linköping University.

Ringius, Lasse. 1997. "Environmental NGOs and Regime Change: The Case of Ocean Dumping of Radioactive Waste." *European Journal of International Relations* 3, March: 61–104.

Rippon, Simon. 1983. "Ocean Disposal." *Nuclear News* 26 (March): 76–79.

Riseborough, R. W. 1970. "The Sea: Should We Now Write It Off as a Future Garbage Pit?" In *No Deposit—No Return*, ed. H. Johnson. Addison-Wesley.

Risse, Thomas. 2000. "'Let's Argue!': Communicative Action in World Politics." *International Organization* 54 (winter): 1–39.

Risse-Kappen, Thomas. 1994. "Ideas Do Not Float Freely: Transnational Coalitions, Domestic Structures, and the End of the Cold War." *International Organization* 48 (spring): 185–214.

Risse-Kappen, Thomas. 1995a. "Bringing Transnational Relations Back In: Introduction." In *Bringing Transnational Relations Back In*, ed. T. Risse-Kappen. Cambridge University Press.

Risse-Kappen, Thomas. 1995b. "Structures of Governance and Transnational Relations: What Have We Learned?" In *Bringing Transnational Relations Bank In*, ed. T. Risse-Kappen. Cambridge University Press.

Risse-Kappen, Thomas. 1996. "Exploring the Nature of the Beast: International Relations Theory and Comparative Policy Analysis Meet the European Union." *Journal of Common Market Studies* 34 (March): 53–80.

Roberts, Leslie. 1982. "Ocean Dumping of Radioactive Waste." *BioScience* 32 (November): 773–776.

Rochlin, Gene I. 1979. *Plutonium, Power, and Politics: International Arrangements for the Disposition of Spent Nuclear Fuel*. University of California Press.

Rosenau, James N. 1993. "Environmental Challenges in a Turbulent World." In *The State and Social Power in Global Environmental Politics*, ed. R. Lipschutz and K. Conca. Columbia University Press.

Rosenbaum, Walter A. 1973. *The Politics of Environmental Concern*. Praeger.

Rowland, Wade. 1973. *The Plot to Save the World: The Life and Times of the Stockholm Conference on the Human Environment*. Clarke, Irwin.

Ruggie, John G. 1983. "International Regimes, Transactions, and Change: Embedded Liberalism in the Postwar Economic Order." In *International Regimes*, ed. S. Krasner. Cornell University Press.

Ruggie, John G. 1998a. "Introduction: What Makes the World Hang Together? Neo-Utilitarianism and the Social Constructivist Challenge." In Ruggie, *Constructing the World Polity: Essays on International Institutionalization*. Routledge.

Ruggie, John G. 1998b. "Interests, Identity, and American Foreign Policy." In Ruggie, *Constructing the World Polity*. Routledge.

Russett, Bruce M., and John D. Sullivan. 1971. "Collective Goods and International Organization." *International Organization* 25: 845–865.

Sabatier, Paul A. 1988. "An Advocacy Coalition Framework of Policy Change and the Role of Policy-Oriented Learning Herein." *Policy Sciences* 21: 129–168.

Sabatier, Paul A. 1991. "Toward Better Theories of the Policy Process." *PS: Political Science and Politics* 24 (June): 144–147.

Samstag, Tony. 1983. "Unions Act to Black Nuclear Dumping." *Times* (London), April 7.

Samstag, Tony. 1985. "Talks on Radioactive Dumping." *Times* (London), September 23.

Sand, Peter H., ed. 1992. *The Effectiveness of International Environmental Agreements: A Survey of Existing Legal Instruments*. Grotius.

Sapolsky, Harvey M. 1990. "The Politics of Risk." *Dædalus* 119 (fall), 83–96.

Schachter, Oscar, and Daniel Serwer. 1971. "Marine Pollution Problems and Remedies." *American Journal of International Law* 65: 84–111.

Scharpf, Fritz W. 1997. *Games Real Actors Play: Actor-Centered Institutionalism in Policy Research*. Westview.

Schelling, Thomas C. 1960, 1980. *The Strategy of Conflict*. Harvard University Press

Schelling, Thomas C. 1978. *Micromotives and Macrobehavior*. Norton.

Schenker, M. S. 1973. "Saving the Dying Sea? The London Dumping Convention on Ocean Dumping." *Cornell International Law Journal* 7: 32–48.

Schneider, Mark, and Paul Teske. 1992. "Toward a Theory of the Political Entrepreneur: Evidence from Local Government." *American Political Science Review* 86 (September): 737–747.

Schneider, Stephen H. 1998. "Kyoto Protocol: The Unfinished Agenda." *Climatic Change* 39 (May): 1–21.

Schoon, Nicholas. 1994. "UK Bows to Ban on Dumping N-waste at Sea." *Guardian* (Manchester), February 18.

Sebenius, James K. 1992. "Challenging Conventional Explanations of International Cooperation: Negotiation Analysis and the Case of Epistemic Communities." *International Organization* 46 (winter): 323–365.

Sen, Amartya K. 1977. "Rational Fools: A Critique of the Behavioral Foundations of Economic Theory." *Philosophy and Public Affairs* 6 (summer): 317–344.

Serwer, Daniel. 1972. "International Co-Operation for Pollution Control." In *Law, Institutions, and the Global Environment.*, ed. J. Hargrove. Oceana.

Shabecoff, Philip. 1982. "Agency May Alter Atom Waste Policy." *New York Times*, January 15.

Shepsle, Kenneth A. 1989. "Studying Institutions. Some Lessons from the Rational Choice Approach." *Journal of Theoretical Politics* 1: 131–149.

Short, Herb. 1984. "Sea Burial of Radwaste: Still Drowned in Debate." *Chemical Engineering* 91 (March 5): 14–18.

Sielen, Alan B. 1988. "Sea Changes? Ocean Dumping and International Regulation." *Georgetown International Environmental Law Review* 1 (spring): 1–32.

Sikkink, Kathryn. 1991. *Ideas and Institutions: Developmentalism in Brazil and Argentina*. Cornell University Press.

Sikkink, Kathryn. 1993. "The Power of Principled Ideas: Human Rights Policies in the United States and Western Europe." In *Ideas and Foreign Policy*, ed. J. Goldstein and R. Keohane. Cornell University Press.

Skolnikoff, Eugene B. 1997. Same Science, Differing Policies; the Saga of Global Climate Change. Report 22, Joint Program on the Science and Policy of Global Change, MIT.

Slater, Jim. 1983. "Radioactive Waste Dumping at Sea" (letter). *Times* (London), August 4.

Slovic, Paul, Mark Layman, and James H. Flynn. 1991. "Risk Perception, Trust, and Nuclear Waste: Lessons from Yucca Mountain." *Environment* 33 (April): 6–11 and 28–30.

Smith, Robert M. 1970. "Panel Urges Curbs on Ocean Dumping; Nixon Hails Report." *New York Times*, October 8.

Snidal, Duncan. 1986. "The Game Theory of International Politics." In *Cooperation under Anarchy*, ed. K. Oye. Princeton University Press.

Snidal, Duncan. 1991. "Relative Gains and the Pattern of International Cooperation." *American Political Science Review* 85 (September): 701–726.

Snidal, Duncan. 1995. "The Politics of Scope: Endogenous Actors, Heterogeneity, and Institutions." In *Local Commons and Global Interdependence*, ed. R. Keohane and E. Ostrom. Sage.

Soroos, Marvin S. 1981. "Trends in the Perception of Ecological Problems in the United Nations General Debates." *Human Ecology* 9 (March): 23–45.

Spiller, Judith, and Cynthia Hayden. 1988. "Radwaste at Sea: A New Era of Polarization or a New Basis for Consensus?" *Ocean Development and International Law* 19: 345–366.

Spiller, Judith, and Alison Rieser. 1986. "Scientific Fact and Value in U.S. Ocean Dumping Policy." *Policy Studies Review* 6 (November): 389–398.

Springer, Allen L. 1988. "United States Environmental Policy and International Law: Stockholm Principle 21 Revisited." In *International Environmental Diplomacy*, ed. J. Carroll. Cambridge University Press.

Stairs, Kevin, and Peter Taylor. 1992. "Non-Governmental Organizations and the Legal Protection of the Oceans: A Case Study." In *The International Politics of the Environment*, ed. A. Hurrell and B. Kingsbury. Oxford University Press.

Stebbing, A. R. D. 1992. "Environmental Capacity and the Precautionary Principle." *Marine Pollution Bulletin* 24 (June): 287–295.

Steinbruner, John D. 1974. *The Cybernetic Theory of Decision.* Princeton University Press.

Stevenson, John. Statement. In U.S. Senate, *Ocean Waste Disposal: March 2, 3; April 15, 21, 22, and 28, 1971* (92nd Congress, 1st Session).

Stokke, Olav S. 1997. "Regimes as Governance Systems." In *Global Governance*, ed. O. Young. MIT Press.

Stokke, Olav S. 1998. "Nuclear Dumping in Arctic Seas: Russian Implementation of the London Convention." In *The Implementation and Effectiveness of International Environmental Commitments*, ed. D. Victor et al. MIT Press.

Strange, Susan. 1983. "*Cave! hic dragones:* A Critique of Regime Analysis." In *International Regimes*, ed. S. Krasner. Cornell University Press.

Straub, Conrad P. 1964. *Low-Level Radioactive Wastes: Their Handling, Treatment, and Disposal.* U.S. Atomic Energy Commission.

Strong, Maurice. 1971. Statement. In United States Senate, Committee on Commerce, and House of Representatives, Committee on Science and Astronautics, *International Environmental Science: May 25 and 26, 1971* (92nd Congress, 1st session).

Tannenwald, Nina. 1999. "The Nuclear Taboo: The United States and the Normative Basis of Nuclear Non-Use." *International Organization* 53 (summer): 433–468.

Templeton, W. L. 1982. "Dumping Packaged Low Level Wastes in the Deep Ocean." *Nuclear Engineering International* 28 (February): 36–41.

Templeton, W. L., and A. Preston. 1982. "Ocean Disposal of Radioactive Wastes." *Radioactive Waste Management and the Nuclear Fuel Cycle* 3 (September): 75–113.

Thomas, Jo. 1984. "Greenpeace Aims at Headlines First." *International Herald Tribune*, September 4.

Timagenis, G. J. 1980. *International Control of Marine Pollution, Volume 1.* Oceana.

Train, Russell. 1972. Statement. In United States Senate, Committee on Foreign Relations, *U.N. Conference on Human Environment—Preparations and Prospects: May 3, 4, and 5, 1972* (92nd Congress, 2nd Session).

Trumbull, Robert. 1980. "Pacific Governors Oppose Dumping Atom Wastes." *New York Times*, October 5.

Trupp, Phil. 1984. "Nuclear Subs to Settle on Dry Land." *Oceans* 17 (July): 34–35.

Tunley, Roul. 1974. "Fresh Start for the Great Lakes." *Reader's Digest* 105 (December): 217–224.

Underdal, Arild. 1987. "International Cooperation: Transforming 'Needs' into 'Deeds.'" *Journal of Peace Research* 24: 167–183.

Underdal, Arild. 1991. "International Cooperation and Political Engineering." In *Global Policy Studies*, ed. S. Nagel. Macmillan.

Underdal, Arild. 1994. "Leadership Theory: Rediscovering the Arts of Management." In *International Multilateral Negotiation*, ed. I. Zartman. Jossey-Bass.

Underdal, Arild. 1995. "The Study of International Regimes." *Journal of Peace Research* 32 (February): 113–119.

Underdal, Arild. 1998. "Explaining Compliance and Defection: Three Models." *European Journal of International Relations* 4 (March): 5–30.

UNEP. 1997. *Register of International Treaties and Other Agreements in the Field of the Environment.* http://unephq.unep.org:70/00/un/unep/elipac/intl_leg/register/reg-intr.txt

UNESCO. 1970. Use and Conservation of the Biosphere: Proceedings of the Intergovernmental Conference of Experts on the Scientific Basis for the Rational Use and Conservation of the Resources of the Biosphere, Paris, September 4–13, 1968.

U.S. Environmental Protection Agency. 1994. Radioactive Waste Disposal: An Environmental Perspective. Report EPA 402-K-94-001.

U.S. General Accounting Office. 1995. Radioactive Waste: Status of Commercial Low-Level Waste Facilities. Report GAO/RCED-95-67.

U.S. House of Representatives, Subcommittee on Fisheries and Wildlife Conservation and the Subcommittee on Oceanography of the Committee on Merchant Marine and Fisheries. 1971. Ocean Dumping of Waste Materials: April 5, 6, 7, 1971 (92nd Congress, 1st session).

U.S. House of Representatives, Committee on Merchant Marine and Fisheries. 1983. *Disposal of Decommissioned Nuclear Submarines: October 19, 1982* (97th Congress, 2nd session).

U.S. House of Representatives, Subcommittee of the Committee on Government Operations. 1983. Ocean Dumping of Radioactive Waste off the Pacific Coast: October 7, 1980 (96th Congress, 2nd session).

U.S. Nuclear Regulatory Commission. 1996. Radioactive Waste: Production, Storage, Disposal. Report NUREG/BR-0216.

U.S. Senate. 1971. Ocean Waste Disposal: March 2, 3; April 15, 21, 22, and 28 (92nd Congress, 1st Session).

Van Dyke, Jon, 1988. "Ocean Disposal of Nuclear Wastes." *Marine Policy* 12 (April): 82–95.

Van Dyke, Jon, Kirk R. Smith, and Suliana Siwatibau. 1984. "Nuclear Activities and the Pacific Islanders." *Energy* 9: 733–750.

van Weers, A. W., B. Verkerk, and C. Koning. 1982. "Sea Disposal Experience of the Netherlands." In proceedings of Symposium on Waste Management, Tucson.

Vartanov, Raphael, and Charles D. Hollister. 1997. "Nuclear Legacy of the Cold War: Russian Policy and Ocean Disposal." *Marine Policy* 21: 1–15.

Waldichuk, Michael. 1982. "An International Perspective on Global Marine Pollution." In *Impact of Marine Pollution on Society*, ed. V. Tippie and D. Kester. Praeger.

Walsh, Annemarie H. 1982. "The Political Context." In *Impact of Marine Pollution on Society*, ed. V. Tippie and D. Kester. Praeger.

Walsh, James P. 1981. "U.S. Policy on Marine Pollution: Changes Ahead." *Oceanus* 24 (spring): 18–24.

Waltz, Kenneth N. 1954. *Man, the State and War: A Theoretical Analysis*. Columbia University Press.

Waltz, Kenneth N. 1979. *Theory of International Politics*. Addison-Wesley. Reprinted in *Neorealism and Its Critics*, ed. R. Keohane (Columbia University Press, 1986).

Wapner, Paul. 1995. "Politics Beyond the State: Environmental Activism and World Civic Politics." *World Politics* 47 (April): 311–340.

Ward, Barbara. 1973. "Speech for Stockholm." In *Who Speaks for Earth?* ed. M. Strong. Norton.

Ward, Barbara, and René Dubos. 1972. *Only One Earth: The Care and Maintenance of a Small Planet*. André Deutsch.

Wassermann, Ursula. 1985a. "Uncontrolled Transport of Nuclear Materials." *Journal of World Trade Law* 19 (March-April): 178–181.

Wassermann, Ursula. 1985b. "Disposal of Radioactive Waste." *Journal of World Trade Law* 19 (July-August): 425–428.

Weart, Spencer R. 1988. *Nuclear Fear: A History of Images*. Harvard University Press.

Weinberg, Alvin. 1972. "Science and Trans-science." *Minerva* 10 (April): 209–222.

Weinstein-Bacal, Stuart. 1978. "The Ocean Dumping Dilemma." *Lawyer of the Americas* 10: 868–920.

Weisman, Steven R. 1991. "Japan Yields to U.S. on Drift Net Fishing: Government, Citing Pressure, Says It Will Halt Practice in North Pacific." *International Herald Tribune*, November 27.

Wendt, Alexander E. 1987. "The Agent-Structure Problem in International Relations Theory." *International Organization* 41 (summer): 335–370.

Wendt, Alexander. 1992. "Anarchy Is What States Make of It: The Social Construction of Power Politics." *International Organization* 46 (spring): 391–425.

Wenk, Edward Jr. 1972. *The Politics of the Ocean*. University of Washington Press.

Whipple, C. G. 1986. "Dealing with Uncertainty about Risk in Risk Management." In *Hazards: Technology and Fairness*. National Academy Press.

Wildavsky, Aaron. 1997. *But Is It True? A Citizen's Guide to Environmental Health and Safety Issues*. Harvard University Press.

Wilson, Edward O. 1994. *Naturalist*. Warner Books.

Wilson, James Q. 1980. "The Politics of Regulation." In *The Politics of Regulation*, ed. J. Wilson. Basic Books.

Wilson, James Q. 1995. "New Politics, New Elites, Old Publics." In *The New Politics of Public Policy*, ed. M. Landy and M. Levin. Johns Hopkins University Press.

Woods, Ngaire. 1995. "Economic Ideas and International Relations: Beyond Rational Neglect." *International Studies Quarterly* 39 (June): 161–180.

Wright, Pearce. 1983a. "Protesters Attack Nuclear Dumping." *Times* (London), February 15.

Wright, Pearce. 1983b. "Britain Defies Ban on Dumping Waste." *Times* (London), February 18.

Yee, Albert S. 1996. "The Causal Effects of Ideas on Policies." *International Organization* 50 (winter): 69–108.

Young, Oran R. 1983. "Regime Dynamics: The Rise and Fall of International Regimes." In *International Regimes*, ed. S. Krasner. Cornell University Press.

Young, Oran R. 1986. "International Regimes: Toward a New Theory of Institutions." *World Politics* 39 (October): 104–122.

Young, Oran R. 1989. "The Politics of International Regime Formation: Managing Natural Resources and the Environment." *International Organization* 43 (summer): 349–375.

Young, Oran R. 1991. "Political Leadership and Regime Formation: On the Development of Institutions in International Society." *International Organization* 45 (summer): 281–308.

Young, Oran R., and Gail Osherenko, eds. 1993a. *Polar Politics: Creating International Environmental Regimes*. Cornell University Press.

Young, Oran R., and Gail Osherenko. 1993b. "International Regime Formation: Findings, Research Priorities, and Applications." In *Polar Politics*, ed. O. Young and G. Osherenko. Cornell University Press.

Interviews Cited

Aarkrog, Asker, Head, Ecology Section, Environmental Science and Technology Department, Risø National Laboratory, Denmark. Interviewed March 20, 1992 in Risø.

Dyer, Robert S., Chief, Environmental Studies Branch, Office of Radiation Programs, U.S. Environmental Protection Agency. Interviewed September 27, 1991 in Washington.

Hansen, Kirsten F., Section Head, National Agency of Environmental Protection, Denmark. Interviewed January 17, 1990 in Hørsholm, Denmark.

Juste Ruiz, José, Spanish delegate to LC meetings. Interviewed November 29, 1991 in London.

Kuramochi, Takao, First Secretary, Embassy of Japan, Washington, DC. Interviewed August 30, 1991 in Washington.

Kuwabara, Sachiko, member, Stockholm secretariat. Interviewed August 26 and 27, 1991 in New York.

Lettow, Charles F., U.S. delegate to LC negotiations (1971–72) and Counsel, U.S. Council on Environmental Quality (1970–1973). Interviewed September 24, 1991 and April 2, 1994, in Washington.

McManus, Robert J., U.S. delegate to LC negotiations (1972), former Director, Oceans Division, Office of International Activities, U.S. Environmental Protection Agency, and former General Counsel, National Oceanic and Atmospheric Administration. Interviewed August 29, 1991 and March 30, 1994 in Washington.

Olsen, Ole V., Oceanographer, Danish Institute for Fisheries and Marine Research, Charlottenlund, Denmark, advisor to Denmark's delegation to the LC negotiations in 1972. Interviewed March 19, 1992 in Charlottenlund.

Sielen, Alan B., Director, Multilateral Staff, Office of International Activities, U.S. Environmental Protection Agency. Interviewed August 29, 1991 and April 4, 1994 in Washington.

Thacher, Peter S., member of secretariat. Interviewed May 2, 1991 in Cambridge, Massachusetts, and August 14, 1991 in Stonington, Connecticut.

Wood-Thomas, Bryan C., Environmental Scientist, Marine Policy Programs, Office of International Activities, U.S. Environmental Protection Agency. Interviewed August 29, 1991 in Washington, November 27, 1991 in London, and March 31, 1994 in Washington.

Index

Advisory Committee on Marine Pollution, 188–190
Agency, 35
Alpha rays, 21–22
Antarctica, 90
Aström, Sverker, 88–89
Atomic Energy Commission, US, 27–28, 79
Axelrod, Robert, 49

Batisse, Michel, 88
Becquerel (unit), 22
Beta rays, 21–22
Biosphere Conference, 68–69, 72, 88, 103
Britain, 4–5, 131–132, 135
 and assimilative capacity, 149
 disposal facilities of, 32
 and global ocean dumping regime, 104, 114
 and global radwaste ban, 31, 151–152
 and radwaste regulation under London Convention, 98, 102, 106–107
 protests against radwaste disposal by, 30–31, 136, 139–148, 153–154
 radwaste disposal by, 24–27, 29
 view of environment in, 91

Camplin, W. C., 159
Canadian International Development Agency, 89
CFCs, 165
Chernobyl incident, 22–23, 57

Chlorinated hydrocarbon, 78, 98
Chlorofluorocarbons, 165
CITES (endangered species trade regime), 167
Climate change, 2, 34, 40, 165–166, 185
Collective good, 40, 48, 52–53, 118. *See also* Public good
Commoner, Barry, 69, 76–77
Contamination, 64, 188
Cooperation, 5–6, 157–158
Council on Environmental Quality, 60–66, 75–76, 80, 117
Cousteau, Jacques, 56, 76–81, 122, 154
Curie (unit), 22

Davis, Jackson, 138
DDT, 60, 69, 82
Deliberation, 37, 191–192
Department of Army, US, 74–75, 85
Department of Defense, US, 151, 155
Department of Energy, US, 155
Department of State, US, 68, 80, 84
Developing countries, 91, 92, 99–106
Dingell, John, 65
Disposal, 22
Diversity, biological, 2, 44, 165
Dubos, René, 127
DuPont, 167

Ehrlich, Anna, 78–79
Ehrlich, Paul, 69, 78–79

Environmental nongovernmental organizations, 4–5, 13, 30, 132, 152, 155–164, 169–170, 174–175, 181, 184, 186, 192
Environmental Protection Agency, US
and assimilative capacity, 65
establishment of, 60–61
and global radwaste-disposal ban, 155
and London Convention, 33
and radwaste disposal 29, 31, 134
Epistemic communities, 8, 36, 53–57, 110–113, 158–159, 169–170, 174, 186, 192
European Commission, 137
European Union, 35, 43, 185

Finger, Matthias, 163–164
Focal points, 51, 120
Focusing events, 11, 85, 165
French Atomic Energy Commission, 76

Gains, relative, 47
Gamma rays, 21–22
GATT, 84
GELTSPAP, 96
General Agreement on Tariffs and Trade, 84
GESAMP, 76–77, 95–99, 159, 161, 188–190
Gonzalez, Felipe, 137, 155
Greenpeace International, 5, 13, 136–139, 142–145, 149–154, 186, 191
Group of Experts on Long-Term Scientific Policy and Planning, 96
Gusfield, Joseph, 125

Haas, Ernst, 52–54, 113
Haas, Peter, 8, 36, 54–57, 110, 112
Haggard, Stephan, 6
Hamilton, E. I., 159
Hegemonic stability, 47, 107, 157
Hegemony, 177
Heyerdahl, Thor, 76–81, 96, 122, 127

High-level radioactive waste, 23, 26, 28, 30, 33, 62, 79, 101–103, 106, 134–135
Hill, H. D., 159
Hirschman, Albert, 40, 177–178
Hollings, Ernest, 77, 81, 121–122
Hollister, Charles, 140
Human Environment Declaration, 104

Ideas
contextual characteristics of, 43–44, 183
economic, 44, 177–178
and effectiveness of environmental regimes, 192–193
and efficiency, 45, 176
and interests, 3, 182, 193
and distributional issues, 44–45, 176, 192
influence of, 3, 43–44, 183
normative, 17
power of, 3, 43–44, 178
public, 1
IGPRAD, 148–150, 160–161
Images, 10, 17, 41, 44, 165, 182
Intergovernmental Maritime Consultative Organization, 68, 70, 93, 96, 99–100
Intergovernmental Panel of Experts on Radioactive Waste Disposal at Sea, 148–150, 160–161
Intergovernmental Working Group on Marine Pollution, 77, 84–85, 90, 92, 95–107, 125, 127
International Atomic Energy Agency, 23, 26, 29, 95, 99–103, 106, 144, 151, 159–160
International Commission on Radiological Protection, 29, 160
International Council of Scientific Unions, 144
International Maritime Organization, 16, 29, 68
International Oceanographic Commission, 95, 99

International Transport Unions Federation, 144
International Union for Conservation of Nature and Natural Resources, 68
Issue networks, 42, 53

Japanese Science and Technology Agency, 133
Joint Group of Experts on Scientific Aspects on Marine Pollution. *See* GESAMP
Justification
 of policy, 8, 9, 37, 174, 183
 as regulatory principle, 190

Kean, Thomas, 135
Kelman, Steven, 41
Keohane, Robert, 49, 118
Kingdon, John, 42
Knowledge brokers, 8–9, 174
Krasner, Stephen, 2

Leaders
 entrepreneurial, 51, 87, 120, 185
 intellectual, 51, 87, 120, 173–174, 185
 structural, 51, 87, 120, 185
Levy, Marc, 16
Limited Test Ban Treaty, 22
Lipschutz, Ronnie, 163
Lists, black and gray, 93–95, 101, 106, 108, 138, 173, 189
Litfin, Karen, 8, 55
London Convention, 5, 15–16, 28, 31, 33, 105–106, 138, 141, 152, 161, 186–192
London Dumping Convention, 140
Low-level radioactive waste, 24, 27

Magnuson, Warren, 178
Majone, Giandomenico, 45, 185
Manhattan Project, 134
Marine Pollution Division, National Agency of Environmental Protection, Denmark, 139

Marine Protection, Research, and Sanctuaries Act (1972), 67, 133
Mead, Margaret, 69
Mediterranean Action Plan, 36–37, 54–56, 112–113
Mediterranean Sea, 8, 54, 76, 166
Metaphors, 10–11, 44, 165, 182, 192
Middle-range theory, 35
Mill tailings, 23–24
Ministry of Defense, UK, 31
Monnet, Jean, 38
Moore, Mark, 42–43
Muskie, Edmund, 61
Mussard, Jean, 88–89

National Academy of Sciences, US, 75
National Advisory Committee on Oceans and Atmosphere, US, 140
National Oceanic and Atmospheric Administration, US, 65, 75, 77, 140, 155
National Union of Railwaymen, UK, 143
National Union of Seamen, UK, 142
NATO, 83
Nelson, Gaylord, 62, 122
Nerve gas, 74, 85
Nixon, Richard, 60–65, 76–77, 114
Norms, 2, 17, 193
 cultural, 177
 embedded, 53
 of global ocean dumping regime, 4, 13, 32, 126, 130, 141, 149, 152, 156, 161, 164
 of hegemon, 177–179
 for ocean protection, 75, 80, 85–86, 122
 of regime, 1–4, 13–14, 48, 168, 171, 175, 192
 setting of, 8, 11, 122
 and transnational entrepreneur coalitions, 39
North Atlantic Treaty Organization, 83
Nuclear Energy Agency, US, 26–29, 132
Nuclear Engineering International, 153
Nuclear weapons, testing of, 22, 165

Oceanic Society, 135
Oceans
 assimilative capacity of, 64–67, 111, 149, 161, 188–189
 Atlantic, 25–29, 63, 74, 76, 114, 132, 136–139, 142, 145, 147, 150, 159
 "dying ," 10–11, 67, 74, 90, 111–112, 115–116, 122–124, 165, 172, 177–179
 Pacific, 25, 28–29, 63, 147
OECD, 83
Office of Management and Budget, US, 63
Oil rigs, dumping from, 165
Organization for Economic Cooperation and Development, 83
Oslo Convention, 31, 71, 99–102, 114
Ozone, 2, 34, 44, 51, 55, 57, 159, 165, 166

Pacific Basin Development Council, 133
Pareto improvements, 45
PCBs, 69, 78, 160–161
Pearson, Charles, 114
Persuasion, 3, 12, 14, 17, 39, 41, 45, 52, 56, 111, 113, 117–118, 121, 128–129, 165, 174, 178, 180
Policy community, 42, 53
Policy entrepreneurs, 11–14, 37–38, 44
Policy streams, 38, 44
Policy subsystem, 42, 53
Policy window, 38–39, 128, 180
Polychlorinated biphenyls, 69, 78, 160–161
Precautionary principle, 149, 161, 188
Preparatory Committee of Stockholm conference, 85, 89–93, 126
Princen, Thomas, 163–164
Public good, 4, 17, 50, 119. *See also* Collective good

Radiation, 21
Radioactivity, 21
Radwaste disposal, 3
Raustiala, Kal, 163

Reagan, Ronald, 135
Regime analysis
 combining approaches to, 184
 interest-based approach to, 6–7, 36, 48–52
 knowledge-based approach to, 6–8, 36–37, 52–57
 power-based approach to, 6–7, 35–36, 45–48
Regimes, 2, 4
 change of, 4–5, 14
 mediating effects of, 13–14, 155–156
 negative effects of, 186
Reich, Robert, 41,
Rio Conference, 148
Ruckelshaus, William, 64
Ruggie, John, 37

"Save the whales," 44, 165
Sealing, 165
Self-help, 46
Ships, oil pollution from, 66, 93, 166
Sielen, Alan, 146
Simmons, Beth, 6
Slogans, 11
"Spaceship Earth," 44
Stockholm Action Plan, 104
Stockholm conference, 18, 68–69, 76, 83–84, 87–88, 91, 103–104, 107–108, 113, 115, 119, 126–127, 172, 180. *See also* UN Conference on Human Environment
Stockholm secretariat, 87–95, 108, 112–113, 117–121, 125–129, 173–174, 177, 180, 187
Storage, 22
Strong, Maurice, 89, 92, 95–96, 113, 125–129, 173, 180
Submarines, 29, 31, 134–135, 140, 148
Symbols, 10–11, 17, 44, 165

TEC approach, 35–45, 175
Thacher, Peter, 126
Thant, U, 89
Thatcher, Margaret, 143, 154
Trade Unions Congress, UK, 143–144

Train, Russell, 59, 114–115
Transnational entrepreneur coalitions, 35–45, 175
Transport and General Workers' Union, UK, 143
Transuranic waste, 23
Two-level games and players, 39, 154

UN Conference on Environment and Development, 148
UN Conference on Human Environment, 18, 62, 68, 76. *See also* Stockholm conference
UN Conference on Law of Sea, 25, 107, 132, 184
UN Convention on Law of Sea, 31
UN Educational, Scientific, and Cultural Organization, 68–70, 88–89, 96, 99
UN Food and Agriculture Organization, 68–72, 89, 95, 99–100
UN Institute for Training and Research, 99
UN Seabed Committee, 74–75, 104
Underdal, Arild, 118, 120
United States, 4–5, 28, 131
  amount of low-level waste produced in, 24
  disposal facilities of, 31–32,
  "dying oceans" idea and, 10–11
  and global radwaste ban, 31, 150–151
  and London Convention, 97–98, 100–106
  protests against radwaste-disposal programs in, 28–30, 134–135, 140–148
  radwaste-disposal programs of, 27–31, 133–134
  and regime change, 156–157
  and regime formation, 7, 10, 80–81, 84, 107, 117–118, 172–179
  reversal of foreign policy on radwaste disposal by, 150–151
  view of assimilative capacity in, 65
  view of environment in, 60, 90–91

Waltz, Kenneth, 45–46
Wapner, Paul, 163
Ward, Barbara, 116, 127
Weapons
  biological, 79, 101–102, 106
  chemical, 17, 71, 79, 101–102, 106
Whaling, 104, 159, 165–168
Wilson, James, 40–43
Woods Hole Oceanographic Institute, 140
World Health Organization, 89, 95, 99
World Meteorological Organization, 68, 95, 99
World Wide Fund for Nature, 167

X rays, 21

Yeltsin, Boris, 150
Young, Oran, 6, 50–52, 119–120, 185